El Niño

El Niño

The Weather Phenomenon
That Changed the World

Ross Couper-Johnston

Hodder & Stoughton

First published in Great Britain in 2000
by Hodder and Stoughton
A division of Hodder Headline

10 9 8 7 6 5 4 3 2 1

British Library Cataloguing in Publication Data
A CIP catalogue record for this title
is available from the British Library.

ISBN 0 340 72838 8

Typeset by Hewer Text Ltd, Edinburgh
Printed and bound in Great Britain by
Mackays of Chatham plc, Chatham, Kent

Hodder and Stoughton
A division of Hodder Headline
338 Euston Road
London NW1 3BH

To Lucy,
whose love and fingerprints are all over this book.

Contents

List of Illustrations

Acknowledgements

I'm deeply indebted to the many scientists and academics that assisted me during the research and writing stages. Several have been considerably generous in welcoming me into their offices, reviewing portions of the text and providing photographs, so many thanks on this account to Rob Allan, Bryant Allen, David Arnold, Wolf Arntz, Peter Ashton, Dick Barber, Steve Bourget, Menno Bouma, Cesar Caviedes, Marty Crump, Mike Davey, Mark Eakin, Tsegay Georgis, Niel Gunson, Brian Hayden, Roger Hewitt, Bob Langdon, Patrick Lehodey, Col Limpus, Valerie Loeb, Barry Lovegrove, Jorge Marcos, Shula Marks, Andrew McKechnie, Neville Nicholls, Sarah O'Hara, Alan Pearce, Lesley Potter, Betty Schreiber, Susan Swan, Fritz Trillmich, Colleen Vogel, Peter Whetton, Polly Wiessner, Martin Wikelski and Jeff Wilkerson, There have also been many others who have given considerable time to answering my questions, including Cheryl Anderson, Enrique Angulo, Chris Ballard, Ian Boyd, Eric Cadier, Julian Caldecott, Ros Cameron, Elizabeth Cashdan, Bob Claxton, Carol Colfer, Pilar Cornejo, John Croxall, Chris Dickman, Michael Dillon, Bronwen Douglas, Michael Dove, Mark Elvin, Georgina Endfield, Terry Erwin, Tim Finan, Ben Finney, Jim Fox, Mickey Glantz, Richard Grove, Shirley Hanrahan, Scott Hatch, Graeme Hammer, Michael Hamnett, David Hiser, Bob Hitchcock, Hiromichi Iwasaki, Mark Jenkin, Billy Kessler, Dick Kimber, Gerrit Knaap, Robert Langdon, Nancy Lewis, Antonio Magalhaes, Nate Mantua, John Marko, Betty Meggers, Chris Meinen, Clive Minton, Chris Mitchell, Kathryn Monk, David Nash, Alan Newsome, Eric Noji, Ben Orlove, Will Osborne, Tim Palmer, Sid Plant, German Poveda, Julian Priddle, Rob Rowan, Mike Ryan, Dan Sandweiss, Bill Sater, Harald Scherm, Gary Sharp, Kevin Smith, Greg Smoak, Chris Surman, Bill Sydeman, Phil Trathan, Sandy Tudhope, Reed Wadley, Pete Waylen, Bertin Webster, David Wilson, Hugh Yorkston, and Leon Zann.

Under a constant bombardment for new references to feed my seemingly never-ending bibliography, I'd like to thank for their assistance and good humour Kwame, Sita, Terry and colleagues at the British Library, Paul and colleagues at the London Natural History Museum General Library, and staff at the State Library of Victoria. I am particularly grateful as well to my editor Roland Philipps and his assistant Roseanne Boyle for their patience, understanding and commitment, and to my agent Patrick Walsh for his untiring enthusiasm and encouragement. And finally a few personal thanks, to my father not only for his tremendous support and assistance with the research for this book but for providing me with the werewithal to write it, to Kim and Cam for their help and hospitality while in Melbourne, to Michael S. and Trevor R. for chasing the answers to some very obscure questions, and to Mary, for her love and kindness which inspired me to write this in the first place.

El Niños

1525—26	1782—84
1531—32	1785—86
1535	1790—93
1539—41	1794—97
1544	1799
1546—47	1802—04
1552—53	1806—07
1558—61	1810
1565	1812
1567—68	1814
1574	1817
1578—79	1819
1581—82	1821
1585	1824—25
1587—89	1827—28
1589—91	1830
1596	1832—33
1600—01	1835—36
1604	1837—39
1607—08	1844—46
1646	1850
1618—19	1852—53
1621	1854—55
1624	1857—59
1630—31	1860
1635	1862
1641	1864
1647	1865—66
1650	1867—69
1652	1877—78
1655	1880
1661	1865
1671	1888—89
1681	1891
1683—84	1896—97
1687—88	1899—1900
1692	1902—03
1694—95	1905—06
1697	1911—12
1701	1914—15
1703—04	1918—19
1707—09	1925—26
1713—14	1930—31
1715—16	1939—41
1718	1957—58
1720	1963
1723	1965—66
1725	1969
1728	1972—73
1731	1976—77
1734	1982—83
1737	1986—87
1744	1991—95
1747—48	1997—98
1751	
1754—55	
1761—62	
1765—66	
1768—69	
1769—70	
1772—73	
1776—78	

El Niño, La Niña Chronology

El Niños and **La Niñas** are disruptions of the ocean-atmosphere system in the tropical Pacific, with each event differing in its strength and its effect on global weather patterns.

El Niños are characterised by unusually warm ocean temperatures in the eastern and central Equatorial Pacific compared to **La Niñas** which are characterised by unusually cold ocean temperatures in the same region.

La Niñas

1872—74
1875—76
1879— 80
1886—87
1889—90
1892—93
1903—04
1908—11
1916—18
1919
1921
1922—23
1924—25
1933—34
1938—39
1942—43
1945—6
1948—49
1949—50
1954—55—56
1964
1967—68
1970—71
1973—74
1975—76
1984—85
1988—89
1995—96
1998—2000

This chronology is not intended to present a definitive list.
(For references see Notes and Sources).

Introduction

I cannot imagine any condition which would cause a ship to founder. I cannot conceive of any vital disaster happening to this vessel.[1]

When Captain Edward J. Smith spoke these ill-fated words, just a few years before the *Titanic* set sail on its maiden voyage across the Atlantic, he believed modern shipbuilding had mastered the elements. Yet there was one factor he had not considered. In 1912, like most sea captains, Smith knew nothing of El Niño. Except for the fishermen of Peru, even the term 'El Niño' was unknown. But within a few weeks an unpredictable weather phenomenon was to cast its shadowy figure over the fortunes of the *Titanic*. Just as it had been doing for thousands, and most probably hundreds of thousands of years, El Niño was set once again to change the course of history.

Each year between ten and fifteen thousand icebergs shear off the glaciers of west Greenland and begin drifting down the Labrador Current towards the Grand Banks of Newfoundland some 2,500 kilometres away. In most years, however, the vast majority get drawn into the Gulf Stream, consigned to eternal anonymity by drifting off north-east and melting in the Stream's warm waters. But the normal passage of icebergs, as with so much of the world's climatic activity, can be completely disrupted by El Niño.[2]

By strengthening the northerly winds of winter that blow the icebergs down the east coast of Canada, El Niño ensures many are driven much further south than usual.[3] The annual average number that cross the 48°N line for instance is less than 500. Rarely does an iceberg drift further south past the tails of the Grand Banks, and in a few years there are none at all. By April 1912, there were well over a thousand.

In those days, the North Atlantic shipping track avoided the usual problem of ice by following the arc of a great circle between Ireland and Newfoundland.[4] The *Titanic*'s route skirted twenty-five miles south of a zone marked, 'Field ice between March and July'[5] on the navigational charts. However, its course did take it through another zone, marked by a faint dotted line and the caution, 'Icebergs have been seen within this line April, May and June'.

By April 1912[6], unknown to passengers and crew, the world had been under the spell of an El Niño for several months, and the world's weather systems had been turned upside down. In scattered locations across the world, El Niño was stamping its trademark of havoc.

The previous September, a flood on the Yangtze River had deluged Shanghai and 59 million acres of the rice provinces of Henan, Hubei and Anhui, drowning 100,000 people, with approximately the same number again dying of hunger. Crops failed owing to lack of rain in large parts of Australia, and in India and Russia where famine followed. In southern Africa, hundreds died in Mozambique and Rhodesia. On the Eastern Cape, where drought coincided with an outbreak of the cattle disease East Coast fever, thousands were forced to migrate far afield in search of work. In the sub-Saharan Sahel, tens of thousands of the Tuareg nomads were forced to migrate south into Niger in search of food. And further east

in Somalia, the year became known as *Haaraama'une,* 'The
Eater of Forbidden Fruit', after people were driven to eat
vermin and livestock not killed by the strict Halal method
required by their Islamic faith. Extended drought in the
Amazon basin led to a great fire which burned out of control
for several months, killing thousands of rubber gatherers.
There was also drought across large parts of South-East Asia,
and when the news of it reached Java, a temporary ban was
imposed on all exports of rice from the island on 'penalty of
imprisonment or forced labour at public works for board and
without wages'.[7] In the United States, a March flood on the
lower Mississippi smashed water-height records and severely
discredited the Army Corps of Engineers' over-reliance on
levees. There were also lesser floods on the Ohio, Salinas and
Roanoke Rivers.

By February, Robert Scott and his team in the Antarctic
were battling persistent daily minimum temperatures of below
−30°F, as much as 20°F lower than the long-term average for
that time of year.[8] Only one year in the sixteen years of
modern records approaches anything like the severity of
1912. By March, the low temperatures began producing 'a
thin layer of woolly crystals'[9] on the running surfaces of their
sledges, reducing progress to a crawl just when severe fatigue
began to take hold. Only days before he succumbed to
exhaustion and the cold, Scott himself wrote: 'Our wreck
is certainly due to this sudden advent of severe weather, which
does not seem to have any satisfactory cause.'

At the same time around Greenland and the Arctic, it had
been the mildest winter since well back into the previous
century, allowing exceptional numbers of icebergs to calve
off. Further south in the mid-North Atlantic, however, the
winter of 1911–12 had been an appalling one, the worst for
over thirty years, with storm after storm causing havoc to
shipping. The schooner *Maggie* took two months to cross

from Portugal, only to be crushed by pack ice off Newfoundland. Another schooner, *Corona*, and the Norwegian sealing ship, *Samson*, had sunk during March gales. The French *Niagara* had been holed twice by icebergs.

By April the worst of the storms was over. But a deadly legacy remained: El Niño had left a flotilla of bergs and field-ice in the path of the North Atlantic winter shipping track. In the exceptional years when the ice drifted all the way south to the shipping lane, mariners expected it to be honeycombed and on the point of disintegration. This particular April, ships reported the icebergs as 'hard as steel and blue in tint'[10], an indication that they contained little air and could travel far. One growler – a small low-lying iceberg – made its way down the US coast to within seventy-five nautical miles east of Virginia's Chesapeake Bay.[11]

If Captain Smith didn't know about the exceptional southerly extent of the icebergs that winter at the time he steamed out of Southampton Harbour on 11 April, he was soon to learn. By the evening of Sunday the 14th, several warnings had been received from ships ahead. The *Rappahannock* had even sustained steering-gear damage from ice, and at 10.30 pm flashed a signal lamp message to the *Titanic*, 'just passed through heavy field ice and several icebergs'.[12]

One message, though, never got through. Only a few miles from the site of the impending crash, the *Californian* found itself surrounded by ice, and the captain decided to halt for the day. But before shutting down for the evening, the radio operator began tapping out the message 'lat 42D 3'N, long 49D 9'W, three large bergs five miles southward of us'[13] to warn the *Titanic*. On the *Titanic*, however, fast approaching its ultimate destination, the radio operator's priority was to send to shore the lucrative messages of the fee-paying passengers. As the warning was being passed to the *Titanic*, the

Californian's radio operator found himself cut off. It was the *Titanic*'s last warning.

Hidden among the floating pack of bergs and field-ice was one small, unremarkable specimen, jutting 50–100 feet high out of the water. It was never to be photographed and was days from melting away into the Gulf Stream. But before it disappeared for ever, it was destined to become the most famous iceberg in history. When the *Titanic* struck it a glancing blow, the iceberg's colossal mass held firm, buckling six of the ship's fifteen watertight compartments with its diamond hard base while showering the starboard deck with ice shavings.

Charles Lightoller, the ship's second officer and the most senior crew member to survive, would remember years later the extraordinary ice scene that confronted him as he balanced on some upturned wreckage.

> Never before or since has there been known to be such quantities of icebergs, growlers, field ice and float ice, stretching down with the Labrador Current. In my fifteen years' experience on the Atlantic I had certainly never seen anything like it, not even in the South Atlantic, when in the old days of sailing ships, we used to sometimes go down to 65° south.[14]

In an era when El Niño has risen from total obscurity to being blamed for every extreme weather-related event on the planet, it is worth emphasising that El Niño did not cause the sinking of the *Titanic*. The fundamental seeds of its demise were much more human. Fooled by a trust in the ship's invincibility, driven by the commercial prize of a fast crossing, and armed with an arrogant belief that any dangers ahead would be spotted from the crow's nest, Captain Smith ordered his ship full steam ahead, come what may – El Niño had simply shifted the goalposts.

El Niño's appearance can be viewed like a theatrical set change, temporarily reshuffling the world's weather systems. Hurricane prone coasts remain calm, while normally tranquil zones are decimated by cyclones. Deserts bloom in the rain, while rainforests wilt in drought. And icebergs materialise out of clear, moonless nights to provide storylines for Hollywood blockbusters.

Only now is science beginning to respect the magnitude of El Niño's power and the scale of its effect. The advance in knowledge of the subject has even been likened to the revolution caused thousands of years ago by the understanding of the annual cycle of the seasons.

For centuries, the artisanal fishermen of Paita on Peru's northern coast had noticed a warm counter-current sweeping down from the north each year. For a few weeks it would override the usual cool, north-flowing waters of the Humboldt Current. By the turn of the nineteenth century, the recurring current had become known as the Corriente del Niño, or just plain El Niño. In the lower case, the term in Spanish means 'the male child', but in its upper-case form, it denotes the 'Christ child' or 'Baby Jesus', a reference to its usual appearance around Christmas. Confusingly, however, this is not the same 'El Niño' as we know it today.

Every few years, this warm counter-current was seen to extend much further south, on these occasions lasting several months and bringing with it exceptionally heavy rains. To the fishermen, these meant barren times because of traumatic reductions in fish catches. To the residents around Piura, inland of Paita, these same months were regarded as times of plenty, '. . . The desert soil is soaked by the heavy downpour,' according to a description of the 1891 El Niño event, 'and within a few weeks the whole country is covered by abundant pasture. The natural increase of flocks is practically doubled and cotton can be grown in places where in other years vegetation seems impossible.'[15] It comes as no surprise that these particular years were known as *años de abundancia*, years of abundance, or *años de aguas*, years of rain.

Not only were desirable commodities such as cotton and livestock plentiful, so too were more unexpected sights. During the 1925–6 El Niño, the travelling American naturalist Robert Cushman Murphy witnessed great numbers of mullets 'swimming in the streets' and, more alarmingly, 'crosses and headstones had toppled by the score, and coffins had been flushed from the graves'.[16] There were more surprises down on the coast, where he observed alligators, bananas, yellow-bellied water snakes and reptiles rafting ashore on platforms of plant matter. The harbours, affected by 'some injurious chemical substance', were 'full of emaciated, vermin-infested sea-birds, which lacked the strength to clamber over the sides of the anchored craft'.

Over time, the years of abundance gradually lost their nineteenth-century image of fertility and plenty. Instead, the term El Niño in Peru has evolved to refer to exceptional years of rain and floods. In global terms, the name El Niño has emerged from almost total obscurity to become synonymous with natural disaster.

Today, El Niño is recognised as the cause of the greatest year-to-year fluctuations in the world's climate. From its epicentre in the tropical Pacific, an El Niño shifts currents, tradewinds, rain-bearing systems and storm tracks around the globe. Shock waves reverberate through the world's food chains and national economies. Just as lightning can be seen, when viewed from space, striking widely separated locations at exactly the same time, an El Niño event can be simultaneously experienced across all the continents. Few people on this planet escape being touched.

Although the term 'event' is often used in describing each

separate appearance of an El Niño, it would be misleading to imply that each El Niño is a discrete, isolated occurrence, disconnected from all other El Niños. Each one is simply an extreme on an oscillating continuum, with every new chapter beginning as another ends. Like a mythological beast that slumbers in its lair, emerging only to feed its voracious appetite, El Niño sleeps but never dies.

1

The Uncontrollable Child
A Short History of El Niño

Surely we have here enough to justify a strong suspicion . . . that
we have waves of drought passing over the earth, that we have an
outside cause for the phenomenon that has puzzled us so long – a
phenomenon which we have every reason to believe is subject to
laws as definite as those which hold the planets in their places, and
the knowledge of which is fairly within our reach.
H.C. Russell, Government Astronomer for New South Wales, 1877[1]

Timothy Richard arrived in China in 1870, part of a small
army of British missionaries. China was considered a prime
target by the Church because, being the most civilised of all
non-Christian countries, its conversion promised to awaken
less enlightened nations. In his posting to the north-eastern
province of Shandong, far from the political centre of Beijing,
Richard intended to spend a comfortable life preaching to the
local upper classes. But then an event was to change his life for
ever. By the end of the century, Richard had become one of
the country's most strident voices for national reform, focused
on lifting the vast majority of the population from its burden
of grinding poverty. The turning point for Richard was the
great El Niño of 1877–8.

The years 1877–8 are known in China as the Great Famine.

For two years, while the south of the country suffered a series of devastating floods, hardly a drop of rain fell in northern and central China. 'The sun glitters red in mid-heaven,' described the Shanxi Governor, Tseng Kuochuan, 'and its scorching terrors blaze abroad the earth. The seed-time having now gone by, and rain having failed, it is useless to think of planting.'[2] By the time the rains of the 1879 summer monsoon arrived, between 9 and 13 million people had died, and another 70 million had been severely affected. In some counties of the north-east, up to 90 per cent of the population disappeared. It was without doubt the world's single worst natural disaster on record.

Yet for those two years, the drought in China was only one entry in a catalogue of global climatic disasters. For over eighteen months, an El Niño had ruled the weather systems in almost every region on every continent. In terms of human life lost, the impact of the 1877–8 El Niño exceeded anything recorded before or since. To one scientist who studied the meteorology of the period, the El Niño of 1877–8 is a 'pathological case'.[3] The disasters that followed in its wake became a political battleground for power and influence and, in Richard's case, for souls.

For some provinces, such as Shandong where Richard was based, the drought actually began with the poor summer monsoon of 1876. Day after day, Richard witnessed crowds flocking to the temples praying for rain. He also saw processions of starving peasants trekking from the countryside into the towns in a forlorn quest for food or work. Richard advertised offers of charity on placards nailed to the gates to several towns, along with the exhortation to 'turn from dead idols to the living God'.[4] To him, not only was it his moral duty to aid the physical suffering, famine relief could also present Christianity 'in such a way that it would commend itself to the conscience of the Chinese as superior to anything they themselves possessed'.[5]

But nothing in Richard's sheltered Baptist upbringing in Wales could have prepared him for the horrors to follow. Violence erupted, and bands of up to 500 thieves ganged together to sack granaries of their rice and millet. When personal food supplies were exhausted, houses were pulled down, not only to sell the timber but to devour the roof's sorghum stalks. In certain regions, up to a half of the houses disappeared this way. By this time, all that people had left to sell were their own families. In December 1877, Richard wrote that it had become the norm for people to 'pull down their houses, sell their wives and daughters, eat roots and carrion, clay and refuse . . . If this were not enough to move one's pity, the sight of men and women lying helpless on the roadside, or of dead torn by hungry dogs and magpies, should do.'[6] By the autumn of 1877, the numbers of dying became so overwhelming that corpses were buried in 'ten thousand man-holes'.

By 1878, to the missionaries' great alarm, human flesh appeared on sale in the markets. However, the supply of corpses failed to keep pace with the desperate demand – even the living were butchered. In the words of one Catholic priest, there were no limits. 'Fathers eat sons and daughters, husbands eat wives and children eat their parents.'[7] But it was Richard's vivid descriptions of the unfolding nightmare that awakened international attention, attracting significant donations and sparking a major foreign relief effort.

Over the centuries, famines and other natural disasters, often associated with El Niño, posed serious threats to China's leaders. Resentment from the famine of the 1640–1 El Niño,[8] for example, coincided with a time of rebellion and weakened leadership, hastening the fall of the Ming Dynasty. Peasants regarded such natural phenomena as a break in the cosmic order of the world, signifying a dynasty's decline. For reasons of self-preservation, therefore, it was imperative for the authorities to rally to the plight of their subjects.

In spite of their best intentions, the imperial authorities faced some insurmountable obstacles. Their provincial officials were both apathetic and corrupt. There were very few railways, and what roads existed were in terrible shape. Where relief was available, distribution was almost impossible, as the chairman of one of the foreign relief committees noted. 'Transport can only be carried on by the banded vigilance of the interested owners of the grain, assisted by the trained bands, or militia, which had been hastily got together. Night travel was out of the question.'[9]

When the monsoon rains arrived in 1879, generating crops for the first time in over two years, relief programmes were scaled down. Yet the El Niño drought had left a lasting impact on the role of foreign powers, and the course of the final decades of the Manchu Dynasty. The relief programmes had helped cement a foreign presence which increased over the following years. China was nearing the end of a long period of political and economic stagnation. The enormous toll of the Great Famine exposed the state's impotence to look after its people and seriously undermined the unshakeable faith in the Manchu Dynasty.

Richard's status and ideas, too, had been transformed. In the beginning he believed that in order to reach the masses he must first 'seek the worthy',[10] or in other words, the educated gentry. After the famine, his focus shifted to improving the lives of the impoverished masses. The theme of his message changed from the sufferings of hell to the hell of suffering. In addition, his profile during the famine had left him with considerable political influence, which he later used with some success in advocating a more Western concept of the world, with Western ideas in health, education and science. And until anti-imperialist anger boiled over in the Boxer Rebellion at the turn of the century, Richard remained one of the most influential foreigners in the country.

Throughout the Great Famine, British involvement in the relief effort remained largely confined to the missionaries and the associated Famine Relief Committee centred in Shanghai. Through their efforts, and the donations that flooded in, the British were able to exhibit a humane and caring approach in the face of such human tragedy. However, at the same time in India, where El Niño was having a similarly devastating influence over its monsoon, the British response towards suffering came under a much sterner examination.

India, like China, expected to experience a shortfall in rains in at least part of the country each year – and severe enough every few years to produce famine. Rarely in its history had the shortfall been so prolonged and widespread as in the El Niño of 1877–8. The drought began in 1876 with a partial failure of the rains in the south of the country, becoming a nation-wide calamity in 1877 when the annual summer monsoon hardly appeared at all in large parts of the north, south and north-western provinces. The winter that followed was extremely severe, the coldest in the north-west in 300 years, killing many from exposure and bringing destructive hail and storms in February. Although parts of southern India received rains in 1878, for most of the parched country, the summer monsoon was again a bitter disappointment.

In the early days of the famine, the British authorities were guided by their official code of conduct in times of famine to 'save every life irrespective of the cost',[11] but as the magnitude of the disaster became apparent, adherence to this principle began to waver. The instructions to one relief worker urged 'that the most severe economy should be practised . . . The embarrassment of debt and the weight of taxation, consequent on the expenditure thereby involved, would soon become more fatal to the country itself.' This new stringency would help make the famine one of the worst in India's history.

The task of relief administration was designated to Sir Richard Temple, the Viceroy of India. He subscribed to the British view, based firmly on the principles of Adam Smith's *Wealth of Nations*, that private trade offered the most efficient means of relief. Any steps taken to interfere with it, such as control of grain prices or the importation of emergency supplies, would only be counter-productive.

Instead, the government provided employment on large-scale public works as an alternative solution. With funds as scarce as food, Temple introduced a series of tough cost-saving measures, which meant wages were appallingly low. By mid–1877 over a million were employed on such schemes in the Madras Presidency alone. Many more again were refused work altogether, having been judged as neither 'entitled to it nor needing it'. During an inspection tour, Temple told one applicant: 'Go home, my friend, and come back to me in three weeks, or a month, when you have eaten the proceeds of your hut, your plough, and your ploughing cattle.'[12] By the time the rejects returned, many were too weak to work.

By 1879, over five and a half million people were dead. The original hope of saving every life, irrespective of cost, had been shattered. However, the terrible event had at least provided the spur to establish a Famine Code, and although the El Niños that followed soon after exacted further misery – nearly five million people died in both of the El Niño induced droughts of 1896–7 and 1899–1900 – a more humane approach to famine relief had evolved.

In human terms, the El Niño was most keenly felt in India and China where it claimed nearly 20 million lives. But its shockwaves resounded around the globe.

On the Pacific island of New Caledonia, colonial resentment was also heightened during the 1877–8 El Niño, this time bubbling over into full-scale conflict. In 1877, hostility was already rife among the Kanaks – the collective term for the different native tribes – towards the French colonial administration, over their treatment and the dispossession of much of their land.

The profitable speculation in beef through the 1870s had prompted the colonial graziers to increase their herds on the island enormously. A small portion of the land had been left to the Kanaks to grow their crops, but it was hardly enough for their needs, and the graziers were not prepared to erect fences to separate the native reserves from their own grazing land.

With the arrival of the 1877–8 El Niño, the subsequent drought affected all crops, including fodder available for the cattle. With their range unrestricted, the hungry cattle roamed extensively through the native reserves in search of their sugarcane and yams, destroying dykes and, at times, ruining several months' subsistence in a day. When Atai, one of the great chiefs, complained to the island's governor about the crop-raiding cattle, he was told to build his own fences. The angry chief later remarked, 'When my taro plants go eating their cattle, then I'll build fences.'[13] Long-standing resentment was bubbling up, and rebellion was in the air.

The straw that broke the camel's back was the kidnap in June 1878 of a tribal woman by a colonist, a relatively minor offence in the context of the islanders' discontentment. Enraged, Kanaks exploded in a series of uncoordinated massacres of whites. The French responded brutally, and within four months the insurrection was suppressed. Around 1,200 Kanak lives were lost, as well as 200 French. The status of the defeated Kanaks descended even further, but a consciousness of national spirit that only emerged much later had been born. Atai, who was killed in the slaughter, achieved legendary status.

Other Pacific islands also felt the force of this exceptional El

Niño. The normally pleasant south-easterly trade winds and relatively cool waters of French Polynesia to the east are not conductive to the creation of cyclones. In 1877, the predominant easterly trades were replaced by westerlies. And in February 1878, when many people were living in temporary huts for the copra-making season, a powerful cyclone ripped through the islands. On Anaa, the Tuuhora church was destroyed, only for it to be later rebuilt, then destroyed twice more in the 1906 and 1983 El Niños.

The El Niño also took its toll in the Americas. The people of north-east Brazil are well attuned to periodic drought, but the 1877–8 event is still unsurpassed in its severity. 'When a Brazilian hears of drought, he immediately thinks of Ceará and of 1877. It is as though that place and those two sevens . . . have become the dramatic synthesis of the great droughts that Brazil has suffered,'[14] said twentieth-century sociologist Gilberto Freyre. In the decades preceding 1879, the largely cattle-based economy of the north-east had gradually been integrated with cotton growing, a much more labour-intensive industry. As a consequence, large numbers of rural workers had been attracted from neighbouring regions. When the great drought hit, the cotton and subsistence crops completely failed and landowners did not or could not accept responsibility for their workers' survival. Tens of thousands migrated in search of work, many choosing to follow tales of a rubber boom in the Amazon. Those seeking government relief flocked to the towns, as described by the writer Rodolfo Theophilo: 'Real animal skeletons, with skins blackened by the dust from the roads and stuck to the bones, held out their hands begging from everyone they met.'[15] In the end, starvation and cholera killed 500,000 people.

Meanwhile much of the USA's Midwest and Mississippi Valley basked in unusually high temperatures, forcing many farmers to harvest by moonlight. In the east, the Roanoke and

James Rivers of North Carolina had their worst floods ever recorded, and in the west, the Sacramento and America Rivers overflowed into large parts of the Californian lowlands. According to a letter published in the London *Times*, water broke through the levees in Sacramento, and 'the safety of the entire city was at one time imperilled'.[16]

In the southern states, the exceptional weather over many months created highly favourable conditions for the breeding of the urban mosquito, *Aedes aegypti*, carrier of yellow fever. Mosquito numbers were already high after the heavy rains of the 1877 summer provided numerous breeding sites. Numbers remained abnormally high over the mild winter, with few freezing nights to kill off over-wintering eggs. Spring came early, with many fruit trees in full flower by March, getting the mosquitoes off to a head start. Between April and June, when mosquito eggs hatch and the larvae mature, rainfall was 50 per cent higher than usual and temperatures were 1.5°C above average. The net result was the country's worst ever yellow fever epidemic, causing over 100,000 cases and an estimated deathtoll of around 18,000.

The city of Memphis, where over 10% of the population died, was the worst hit. 'Memphis is today the leper among American cities,' recorded a local newspaper, 'shunned and avoided by its own people, held as a thing of horror, not to be seen or touched, a dreaded spot, fenced out by the surrounding world, to be ravaged by its own pestilence.'[17] The scourge was also described as 'one of the very worst catastrophes in American urban history, its toll in human life surpassing that of the Chicago fire, San Francisco earthquake, and Jonestown flood combined'.[18] Saddled with massive debt and the withdrawal of investment, Memphis's finances were placed under receivership, and the city's charter was revoked. Of the 6,000 white citizens who remained, virtually all contracted the disease and 70 per cent died, leaving an overwhelmingly black population. The change in demographics forced the police to quietly overlook its colour barrier to the recruitment of black citizens who had shown markedly greater resistance to the disease. Over the next few years, with the weather following a more normal pattern and yellow fever ceasing to be a major problem, most new recruits were stripped of their badges.

At the same time in Africa, accounts from several regions suggest the El Niño affected virtually the whole continent. In Morocco, the resident correspondent for the *Jewish World* newspaper lamented: 'If you could see the terrible scenes of misery – poor starving mothers, breaking and pounding up bones they find in the streets, and giving them to their famished children – it would make your heart ache.'[19]

To the east, the annual Nile flood was two metres below normal, leaving at least a third of the valley without water or a replenishing cover of silt. Thousands were forced to migrate from their ancestral homes. Closer to the headwaters of the White Nile, in what today is southern Sudan, the Russian explorer Wilhelm Junker described scenes that haunted him for the rest of his days:

> The numbers of skeletons and human bones lying near the road on the march through the valley had already shown what had taken place there. Death had reaped a rich harvest . . . Starvation could be plainly read in many of the completely mummified bodies. The poor Negroes were literally reduced to skin and bone, and the skin had been tanned to leather by the tropical sun. The hyaenas and vultures must have left a plentiful repast, for they had left a number of corpses un-touched.[20]

In East Africa, rains were so heavy that Lake Tanganyika rose higher than at any point in the previous 200 years. The

Lukuga River, which had previously flowed into the lake, reversed its flow to become a discharge channel for the lake water into the Congo River.

Continuing south, the Western Cape region experienced two years of heavy rains, in places so extreme that they caused disastrous floods. In the east and interior portions of South Africa, against the backdrop of wars against the Zulus and the Xhosa, as we shall see later, severe drought reigned. One resident likened its impact to the fearsome reputation of Xhosa tribes, declaring them 'incapable at their worst of inflicting a tenth part of the injury on the country which has been caused by the lack of rain'.[21]

North in Borneo, the effects of the drought were equally drastic. Norwegian naturalist Carl Bock recorded in his expedition diary of 1879 Dayak descriptions of the previous year's famine: 'Numbers of them had subsisted for more than a year on roots and wild fruits: and even those would have failed had the drought lasted much longer.'[22] Where local leaders were unable to provide relief, or the fruits of the forest proved insufficient, there were large migration movements. The Galalerese of the island of Halmahera migrated to neighbouring islands, and the Hulu Sengai of south-east Borneo emigrated as far away as the Malay Peninsula. In most cases, the people did not return.

Chinese merchants took advantage of the drought conditions in Borneo by persuading coastal Dayaks, who had experienced two years of complete crop failure, to trek into the island's interior to harvest gutta-percha, a latex of dipterocarp trees. Gutta-percha had become a valuable commodity during the second half of the nineteenth century, particularly as insulation for cables. But its harvest in the Borneo interior was often dangerous because it required travel into the territory of potentially warsome tribes, forcing Dayaks to assemble into large groups before following 'unbeaten paths, far from their own territories from which the trees had already been stripped'.[23]

Meanwhile, much further south, large parts of Australia were also in drought, and rivers were drying up. The usually free-flowing Namoi, for example, 'became a series of muddy waterholes'.[24] According to the swagman Joseph Jenkins, not long off the boat from Wales in his escape from an unhappy marriage: 'It is a sorry sight to witness the cattle suffering from want of both food and water. It is the most severe drought within the memory of the Colony's inhabitants.'[25] He made that observation in April 1879. Yet within a few days Joseph was describing torrential rains and by July he wrote in his diary: 'It has rained heavily and all the dams are full. The people begin already to complain of the wet weather.' Finally, a La Niña had arrived. In most El Niño affected countries, the La Niña meant a change in fortunes. Floods abated, and rains brought the first crops in drought-ravaged lands. Yet for regions such as Java and parts of Australia, the La Niña replaced drought with flood as the new scourge of the land, and delivered a third year without crops.

In the unfolding story of the El Niño phenomenon, the period 1877–8 was also significant in other ways. An Australian, better known for another achievement, made the first observation to connect all this strange climatic behaviour.

Historical coincidences of natural disasters and climatic abnormalities between remote locations may have been suggested before the late nineteenth century, but the linkages were never proved. Alexander Beatson, for example, an East India Company representative in the tiny mid-Atlantic outpost of St Helena, had inadvertently stumbled upon the El Niño connection when he wrote: 'The severe drought felt here in 1791 and 1792 was far more calamitous in India . . . [where] owing to a failure of rain, during the above two years, one half of the inhabitants in the northern circars [states] had perished by famine.'[26] Although Beatson did not know it, his observation was made possible by the same network of British colonial outposts that would later begin to piece together the El Niño jigsaw.

When the very first posting of Director-General of the Indian Meteorological Department was granted to the Englishman Henry Blanford in 1875, the most vexing mystery for anyone given his job was to explain why the Indian monsoons would occasionally but spectacularly fail. Only two years into his posting, Blanford was able to witness for himself the full horror of a total monsoon failure. For the British Civil Service, shouldering the responsibility for famine relief, the starvation of over six million Indians meant enormous shame, financial cost and loss of authority. Fully aware of how important it was to human life to comprehend India's baffling meteorological cycles, Blanford turned to a study of conditions beyond India's shores. The best information he could lay his hands on came from Britain's other colonies.

In 1877 one of Blanford's requests for information reached South Australian Government Astronomer Charles Todd, already renowned for his role as chief engineer on the pioneering overland telegraph route between Adelaide and Darwin. In what is probably the first definitive recognition of an international climatic connection, Todd responded: 'Comparing our records [Australia's] with those of India, I find a close correspondence or similarity of seasons with regard to the prevalence of drought, and there can be little or no doubt that severe droughts occur as a rule simultaneously over the two countries.'[27]

Following up this apparent connection between Indian and Australian conditions was only one of a number of leads that fell to Gilbert Walker, one of Blanford's successors at the Indian Meteorological Department. In an unusual career,

Walker explored mathematical relationships in areas as diverse as sports, boomerang physics, bird flight and flute design. But it was his success in establishing statistical correlations between different weather systems that would earn him a lasting reputation. His initial attempts were hampered by the sheer volume of data that had to be processed by hand, but his luck changed with the advent of World War I. Few industries outside the war machinery could expect to have their manpower increased, but Walker found many of the Department's senior scientific staff were drafted off to Europe, and the few that remained were preoccupied with war-related matters. In an instant, dozens of eager Indian civil servants normally assigned to other staff were free to work for him. It was just the break he was looking for.

Spurred on by the observation of Swedish meteorologist Hugo Hildebrand Hildebrandsson of an inverse relationship between the atmospheric pressure of Buenos Aires and Sydney, Walker turned his new army of workers to the business of number crunching. Foremost among them was his faithful assistant, Hem Raj, who possessed a photographic memory for weather charts. From the mass of computations, Walker described three apparent 'seesaws' in atmospheric pressure and rainfall across the world's oceans, where a decrease in one region was matched by the opposite increase in another. The first two seesaws – the North Atlantic Oscillation, between the regions of the Azores and Iceland, and the North Pacific Oscillation, between the Hawaiian Islands and Alaska – held no clues to the mystery of the Indian monsoon variability. However, the third looked more promising. Walker noticed that every few years, the usual atmospheric conditions across the Pacific reversed. With its fulcrum roughly positioned on the International Date Line, the usual high pressure in the east dropped concurrently as the usual low pressure in the west increased. To distinguish it from the other two climatic cycles, Walker coined it the Southern Oscillation.

At first, the discovery of this atmospheric cycle showed great promise in long-range forecasting, but without an explanation of the physical mechanisms to underpin it, interest waned. It would be another fifty years, at the very end of Walker's life, before an explanation was uncovered. Once again, it would need the trigger of another big El Niño.

In 1957–8, the United Nations began a new trend of designating particular themes to years, kicking off with the somewhat unglamorous choice of International Geophysical Year. As a result, new funds allowed a series of moorings to be dropped around the world's oceans to monitor conditions. By coincidence, 1957–8 was one of the strongest El Niños of the century. For the first time, scientists were provided with data on what was happening to the seas and winds far out into the Pacific. To the surprise of the scientists, the information collected by the weather-buoys showed a very different picture to the one that was expected. For the first time during an El Niño, it was possible to see conditions way beyond the realms of the Peruvian fishermen.

Among the small band of scientists attempting to make sense of the 1957–8 data, the Norwegian-American meteorologist Jacob Bjerknes could not have had a better pedigree. As a young man, he had collaborated with his father Wilhelm, one of the founders of modern weather forecasting. By the time he was thirty, Jacob had achieved scientific recognition in his own right for his work on the life cycles of air masses and tropical storms. After a gap of almost fifty years, he would again leave his mark on meteorology.

With the 1957 data at his fingertips, Bjerknes noticed that just as Peru was experiencing the coastal flooding and warm sea temperatures of an El Niño year, the Pacific was simultaneously experiencing weakened trade winds. At the same time, there was heavy rain in what was known as the Pacific dry zone, and the atmospheric seesaw of the Southern Oscillation had flipped.

Although no component of Bjerknes's theory was completely new, his genius was to combine the elements together into a coherent framework, underpinned by a solid physical mechanism. Without more data to support his theory, Bjerknes was in some respects stabbing in the dark, yet subsequent research has largely confirmed the fundamentals of his theory. Linking the observations of Peruvian fishermen to those of India-based Gilbert Walker, Bjerknes combined the oceanic El Niño component with the atmospheric Southern Oscillation component. For the first time, the phenomenon was shown to be a coupled process of atmospheric and oceanic changes. An important new name was struck, the El Niño Southern Oscillation, or ENSO for short. (Although strictly speaking the term El Niño refers to just one phase of the overall ENSO system, this book will often use the term 'El Niño' in its broader and more popular sense, implying the system as a whole.)

Although Bjerknes first published his all-encompassing theory in 1969, it was the 1972–3 El Niño that would light the slow fuse of international awareness.[28] There are climatic disasters and related food shortages in parts of the world at any point in time. Normally the causes of these are unconnected, but in 1972 there was an unprecedented chain of both across several continents.

Kenya, southern Brazil and the western coast of South America all suffered severe flooding, while China, Russia, parts of the Middle East, the Indian sub-continent, Australia, Ethiopia and central America all experienced drought. Little rain had fallen in the Sahel zone of northern Africa since 1968, but 1972 was drier still. Between one and two million people starved and over four million cattle died. In 1973, the

Mississippi flood was so severe that the Army Corps of Engineers nearly lost their ongoing battle to stop the river forging a permanent new route down the shorter and steeper exit of the Atchafalaya River, thereby bypassing New Orleans and Baton Rouge altogether.

Although many in the scientific community grasped that the 1972–3 El Niño's many and varied natural disasters held a causal relationship centred on physical events in the waters of the Pacific, the international media did not. For another decade, the meteorological significance of El Niño was unknown outside of Peru.

By mid–1982, some considerable progress had been made in unravelling the mechanisms behind El Niño's idiosyncratic clock. Although there was no agreement over what made it tick, the basic components were understood, and there was some confidence in predicting when the cuckoo would next appear.

If the events of 1972–3 hadn't made the world sit up and take notice, the events a decade later certainly did. A multitude of climatic records was broken, and communities across the world were subjected to some of the worst natural disasters in living memory.

In South America, parts of Ecuador, Peru, southern Brazil, Bolivia, Argentina and Paraguay experienced heavy flooding, while north-east Brazil baked in drought. Further north, Mexico suffered drought, southern California was belted with storms and mudslides and the Gulf of Mexico states were flooded. A series of hailstorms in Texas cost cotton farmers £1.4 billion. Meanwhile, the east coast basked in the warmest winter for twenty-five years and enjoyed fewer hurricanes than in any season last century. Across the Atlantic, floods occurred in southern and central Europe, with several major rivers, including the Loire, Seine, Rhone, Rhine and Mosel, overflowing their banks. In Africa, Zimbabwe, Zambia, Botswana,

Mozambique, Angola, South Africa and Lesotho all witnessed drought. The Sahel, already experiencing long-term below-average rains begun in 1968, suffered its worst phase since 1972–3. Across Asia, a late start and early finish of the monsoon caused drought in southern India, Sri Lanka, Indo-nesia and the Philippines, while high death tolls accompanied floods on the Yangtze River and in Nagasaki. In Thailand there were heavy floods, with parts of Bangkok inundated for four months. In Australia, the rural economy was savaged by the biggest drought since the turn of the twentieth century, while the south-east of the country suffered devastating fires. Fires in Kalimantan were the biggest in history, and continued to burn through shallow coal seams for years to come. In the Pacific, many island groups, including Hawaii, experienced no rain, while other islands closer to the equator, such as Tonga and the Galápagos, suffered exceptional rains. French Poly-nesia was also belted by six separated cyclones, the first time any had struck for seventy-five years. Hurricane Iwa hit Hawaii, the first to do so in twenty-three years.

In terms of international awareness, 1982–3 was to be the turning point for El Niño. As each new flood or typhoon hit, and the length and severity of droughts tightened the screws on many countries, the international media started gradually to make the connection. The term 'El Niño' began a slow introduction into the vocabulary and consciousness of the international public.

After the record-breaking event of 1982–3, a much weaker one followed in 1986–7, followed again by a weak but exceptionally long series of El Niño years between 1990 and 1995. What happened next took everyone by surprise. Scientists had predicted it would be a hundred years before an event on the scale of 1982–3 would strike again, but in 1997–8, the world was hit by another El Niño of comparable strength.

For anyone keeping half an eye on the news in 1997 and 1998, it was hard to miss the term 'El Niño'. Fires in Indonesia, famine in Papua New Guinea, hurricanes in Mexico, record low-water levels in the Panama Canal, floods in Peru and Kenya, drought in North China, unprecedented snow in Tibet, and an explosive tornado season in America – these were just a few of the stories to hit the headlines, all citing El Niño as a major contributory cause. With less validity, El Niño was also touted as a partial excuse for such diverse phenomena as the Asian economic crisis and the temporary decline in the international art market. It even made appearances in parliamentary speeches, Oscar ceremonies, stock fluctuations and TV chat-show gags.

In less than a hundred years, the term El Niño has evolved from its humble beginnings, denoting a minor change in a local current, to a major climatic switch of global proportions. Its connotation has also been transformed from one of benevolence and abundance to one of chaos and destruction. And its notoriety has spread from the fishermen of Peru's northern ports to international media headlines. Its news value in the 1990s rivalled such topics as Viagra and Princess Diana, its mere connection to a story sufficient to make it newsworthy.

But all the time El Niño has been hogging the limelight, La Niña, the less demonstrative little sister, has remained hidden in the shadows. She emerged in 1998–9, clinging to the coat tails of her big brother, to reveal the other half of a troubled, dysfunctional family. El Niño and La Niña are in essence just opposite phases of the same oscillation, but the relationship between the two is not always one of equals.

In general, La Niña's impacts on any particular region are the opposite in nature to those of El Niño, bringing rain where El Niño brings drought, and so on. And La Niña's impacts are usually less devastating. The 1988 La Niña, for example, the strongest for forty years, included in its wake a severe drought over the Great Plains of North America, the inundation of three-quarters of Khartoum, and an exceptionally severe hurricane season in the Atlantic, including the destructive Hurricane Gilbert. Nonetheless, it was only ever categorised as 'moderate' in strength, and its global bill of damages does not in any way compare to that of the 1982–3 or 1997–8 El Niños.

Historically, awareness of La Niña got off to a slow start because along the Peruvian coast where El Niño was first recognised, La Niña's effect was limited to exacerbating the already desert-like conditions, while El Niño's more dramatic impact turned desert into garden, and dry-river beds into raging torrents.

Even today La Niña's effect in many parts of the tropics tends to be perceived as little more than an extension of the norm. Where the success of annual crops is dependent upon monsoonal rains, an increase in precipitation – which goes hand in hand with La Niñas in many countries – is much easier to cope with than rain failure. Furthermore, in what may be nothing more than a geographical quirk, the regions where La Niñas spell drought, such as the mid-tropical Pacific, do not tend to be widely inhabited.

While the big brother invariably flexes greater muscle than his little sister, the two are seldom far apart, with one tending to make an appearance directly after the other. It is a pattern that has been long recognised in the mythology of Australia's Aboriginals. In a story told by a tribe in the south-east of the country, Tiddalick, a giant frog, swallowed all the water in the world leaving the land baked in drought. Pleading for water, all the animals gathered before him, dancing and playing pranks. But Tiddalick just stared implacably ahead with his big eyes and swollen cheeks. Finally, after an eel began dancing on its tail, the frog couldn't hold back any longer, and disgorged all the water, flooding the land.[29]

Within a few decades of European settlement, as noted by the editors of the *MacCulloch Dictionary* in 1841, there was already a recognition of the tendency for a sudden shift from drought to flood: 'Once in such cycles, a year of unmitigated drought prevails, during which no rain falls, and the effects of which are equally intense on the coast, and in the interior. Close upon this visitation follows a year of flood.'[30]

Of all natural disasters, floods are the greatest killers. Their combination with drought, however, is especially deadly. Still reeling from a recent drought, the land may be exceptionally prone to flood erosion, insect plagues may erupt from a reduction in predator numbers, and farmers risk a second year in a row without a successful harvest.

Although the sudden change from one extreme to another can in itself cause damage, the switch is rarely so violent that the land and people affected cannot cope. In fact, if the oscillation between the two states is neither too sudden nor too extreme, then the change can act as nature's balance of good and bad years. For a community struggling after a poor rainy season one year, prolonged rains the next year may yield a much-needed crop bonanza and a return from debt.

In whatever way La Niña and El Niño events manifest themselves on global weather systems, together they are a powerful combination.

It may be worth pondering why the double trouble of this climatic cycle was not identified sooner. First, there were several hurdles to overcome. Until the last century, the cycles of day and night, and the seasons of the year, were believed to be the only predictable variation in the earth's climate. On any longer term scale, climate was thought to be a static, unchanging system. Any variation was considered random

and completely unpredictable, an act of God in a universe
without pattern.

Although evidence for climatic change had been building
up for many years, it was only when Swiss naturalist Louis
Agassiz spent his 1836 summer holidays scratching around
the rocks of the Rhone Valley that a scientifically credible
explanation was proposed. Agassiz's theory of a prehistoric
ice-age covering the world from the North Pole to the edge of
the Mediterranean was accepted relatively quickly within the
fields of natural science.

After that, the floodgates were open. Over the next few
decades, more than a hundred different cycles of periodic
change were proposed. Most of them remain unsubstantiated,
but evidence from instrumental and archaeological records
shows that, without doubt, climatic fluctuations exist on
many different time-scales. European historical records provide proof of freezing conditions between the fourteenth and
nineteenth centuries known by meteorologists as the Little Ice
Age. This unusually cold period saw skating on the River
Thames in London and led to the destruction of civilisation on
Greenland, prompting an early realisation of climatic change
on a century-long scale. Variation on a decadal scale too, such
as the recent examples of the American Dust Bowl years of the
1930s, and the Saharan Sahel drought of 1968 to 1973, have
been well substantiated.

Awareness of periodic fluctuation on a year-to-year scale,
however, proved much more elusive. Such variation was
much easier to dismiss as the randomness of nature's vicissitudes. Seeking an explanation, scientists and historians
looked to rhythms in the activity of the sun, moon and
planets. Theories of various natural laws abounded. Even
by early last century, over a hundred different cycles had been
suggested, none of which is accepted today by mainstream
meteorologists. All had a set interval at their heart, each cycle

with a simple and predictable rhythm. Certainly, no one was looking for a cycle that would peak and trough with apparent randomness at any interval between a year and a decade, and whose normal duration of around twelve months could extend upwards in exceptional cases to several years.

It was not just El Niño's irregular cycle that threw the hounds off the scent. Each event is different from the last, each showing a new face and a new facet of El Niño's personality. For the discovery of the cycle to be extracted from such seemingly chaotic behaviour, and for its physical processes to be reasonably understood, it is no wonder that science had to wait until the late twentieth century. Only now with the information available from satellites and ocean-based monitoring equipment, together with advances in computer technology, has it been possible to simulate such complex processes.

Now that El Niño's signature has been recognised, it seems that El Niño itself has become more visible. Like a manic-depressive prematurely taken off lithium, El Niño's mood swings have recently begun to dip lower and for longer than at any other time this century. For reasons to be discussed in the final chapter, the last twenty years have seen two extraordinarily severe events, as well as one exceptionally long one.

Television may have introduced El Niño to the world, but it is now the Internet that is helping those intrigued to discover more about it. Web-cams in key locations such as aboard research ships and inside storm systems send instantaneous pictures, tracking El Niño's movements like a detective trailing a fugitive. Changes in sea surface temperatures are picked up by satellites and downloaded on to daily maps of the Pacific, with the coloured fingers of the different temperatures evolving over time like the northern lights. For those making their own observations, newsgroups exist for the latest updates on unusual phenomena, including outbreaks of disease and sightings of birds outside their normal range.

In many respects, El Niño is the perfect natural phenomenon for the early twenty-first century. Its impacts provide violent, dramatic, visually strong images perfect for the television age. Surges of flood water cascading through villages, emaciated children staring forlornly into the camera, home footage of hurricane winds uplifting the immovable, and silhouettes of suburban houses backlit by a wall of wildfire, are all favoured currency of the modern broadcaster. In an era of CNN television, the immediacy and drama of natural disasters unfold on screens around the world. Yet its meteoric rise to prominence belies a hidden past that has until now concealed a profound influence on world history, culture and biological evolution.

El Niño has clearly arrived on the world stage. But while it has belatedly achieved widespread awareness, few people still understand what the phenomenon is all about.

2

The Global Connection
An Explanation of the Physical Processes

Acting in concert, the ocean and the atmosphere are capable of music that neither can produce on its own. El Niño is an example of such music.

George Philander, oceanographer:[1]

On 10 September 1982, the research vessel *Conrad* steamed out of Hawaii carrying a team from the Marine Laboratory of Duke University.[2] Its mission was to study the nutrients and plankton life of the Pacific's prevailing equatorial currents. Prevailing, that is, because as they understood it, 1982 definitely wasn't an El Niño year. All the usual indicators such as sea-surface temperatures were showing situation normal, and what is more, if an El Niño was occurring, it would have showed its face months before.

It was the Fijian cooks who first noticed something strange. Normally the night's meal was caught with ease by simply dropping a line off the stern. But when the *Conrad* reached the equator, the cooks could not hook a thing. With frustration growing, one of the ship's engines broke down, and the crew feared they would drift westwards, causing the research trip to fall behind schedule. Only when their position was checked did they realise that they were being carried in the opposite direction, and heading for Peru.

The *Conrad*'s crew were not alone in their ignorance that an El Niño had been under way for several weeks. In many respects, the 1982–3 El Niño did not fit the classic profile of an El Niño. Its development had started much later in the year than normal, and the changes first appeared in the west of the Pacific rather than the east. And to add further confusion, particulate matter in the atmosphere from the eruption of the Mexican volcano El Chichón seven months earlier was masking changes to the sea-surface temperatures. So although the Pacific was in fact betraying several clues that massive changes were taking place even while the *Conrad* was sailing, few people were able to interpret them correctly.

Even as late as October 1982 one of the most respected 'Niñologists', Klaus Wyrtki, pronounced categorically at a conference that there was no El Niño at the time, nor would there be one the following year either. Wyrtki based his assessment on the non-appearance of anomalous warm waters off the Peruvian coast, considered by many scientists in the early 1980s to be the crucial indicator of the onset of an El Niño. However, a few, such as the Australian meteorologist Neville Nicholls, attached more significance to the sudden mid-year switch in the Southern Oscillation. Nicholls was so convinced that the rainless months of another El Niño were headed his way that, contrary to his wife's instructions, he postponed establishing a new garden in their recently purchased home in Melbourne.

Wyrtki's mistake was to assume that the next El Niño was going to evolve and behave as in previous events. This assumption was common among scientists of the time seeking to understand the phenomenon. But if there is one golden rule about El Niño behaviour that has emerged since then, it is that no two events are ever the same. Despite El Niño's ability to surprise and confound scientists with each new appearance, enormous advances have been made over the last three decades in understanding the basic physical processes behind it. An explanation of these processes first needs a brief overview of the global climate.

At first glance, a map of the world's main currents and wind systems might appear to be a twisted mess of interconnecting arrows, as if a heap of iron filings had been tossed near a series of rotating magnets. Underlying this seemingly random muddle, however, is the one fundamental process that drives the earth's climate – the movement of heat from the tropics to the poles. In any real sense, the poles are no further away from the sun than the tropics, yet because of their oblique angle to the sun's rays, and because of the reflectivity of snow and ice, the amount of heat gained from the sun is considerably less at the poles. The oceans and atmosphere are therefore in constant motion redistributing the energy imbalance.

The first stage of this energy redistribution involves warm, moist air rising near the equator, then moving polewards in the upper atmosphere before descending as cool, dry air around 25–30° of latitude on both sides of the equator. It is no coincidence that many of the world's great deserts are situated around this latitude. At the earth's surface, winds return to the equator, driven by a pressure gradient between the high pressure created by the subsidence of the cool, dry air on the one hand, and the low pressure created by the rising, warm, less dense air on the other. This simple convective cycle is known as a Hadley Cell.

The rotation of the earth complicates this picture considerably. Air travelling away from the equator in the upper atmosphere carries with it a 'memory' of the speed from where it originated. Because air at the equator is rotating west to east faster than at higher latitudes, winds with a 'tropical memory' bend to the east to become westerlies – hence the general movement of weather systems at the mid-latitudes from west to east. Likewise, the surface winds travelling towards the equator carry a 'memory' of the slower speeds of their mid-latitude origins, and are overtaken by the higher speed of the earth's rotation closer to the equator. In this way, instead of blowing at right angles to the equator, they travel diagonally towards the west. These are the trade winds, so called because the Spanish used them on their lucrative route from their South American colonies to the Spice Islands.

In the northern hemisphere, the trades operate as north-easterlies, and in the southern hemisphere, as south-easterlies. The two systems of trade winds meet at the intertropical convergence zones, or ITCZs. Each ITCZ migrates north and south with the seasons, its movement controlling the timing of the annual monsoons. Water vapour packed with the sun's converted energy is blown into these zones by the trades, accumulating into vast towering banks of cumulonimbus clouds that extend as high as fifteen kilometres above the earth's surface. Much of the latent energy is converted into the upper atmospheric winds of the Hadley Cell, releasing enormous quantities of rain, and making these regions the wettest on earth. Roughly 40 per cent of global precipitation falls within fifteen degrees of the equator and most of that within three main convergence zones – centred over the Amazon, the Congo and the western Pacific.

The effect of the continuous westward trade winds on each of the great tropical oceans is to drag the surface water warmed by the sun towards the west. Several important differences between the west and the east occur as a consequence. For a start, this results in the oceans in the west being warmer and the sea-levels higher. The sea-level, for instance, is slightly higher on Panama's east coast than it is on its west. And because surface water temperature is conveyed readily to the overlying atmosphere, atmospheric pressure differences are also created. High pressure zones form in the east over the bands of cold water, above which the chilled air is too dense to rise enough for water vapour to condense and form rainclouds. Low pressure zones form in the west above warm pools of water, where the moist air is warmed, expands and rises, eventually cooling and condensing into rainclouds in the upper atmosphere. And the pressure gradient between the two zones drives the trades from the strong high pressure zone of the east to the low of the west.

Below the ocean surface, there are distinct east-west disparities too. The thermocline, the relatively sharp boundary between warm surface waters and the colder, denser water underneath, is also affected. Because of the greater volume of warmer water dragged into the west, the thermocline is angled downwards towards the west. One major result of this is the rising, or upwelling, of cold water along the coastlines of the three great ocean basins. Here, the pull of the trade winds and associated currents drags warmer surface water away from the coast, in turn pulling water up to replace it from below. Because this is also the region where the thermocline is shallowest, it is cold water that is entrained to the surface. Importantly to the economies of those countries closest to these upwellings, a vast store of nutrients is also brought to the surface. The food chain that this supports makes the five major upwellings – off Peru, California, Namibia, Mauritania and the Horn of Africa – the most productive fishing waters in the world.

The overall picture of the atmosphere and oceans is one of a self-sustaining cycle. Each component is linked to another. The distribution of cold and warm water determines the pressure differences, and hence the winds. The winds in turn influence the angle of the thermocline and degree of upwelling. The upwelling reinforces both the temperature and the pressure gradient, and so on. The result of this cycle is the perpetuation of east-west disparities.

Understandably, the most extreme case of this east-west dipole can be found in the largest ocean of all, the Pacific. Compared to the eastern side, where the thin tongue of cold upwelled water stretches almost as far as the Dateline, the warm pool in the west, centred roughly north of New Guinea and east of the Philippines, is 7°–10°C warmer and the sea-level 60 centimetres higher. In addition, the thermocline in the west may be depressed to a maximum depth of 200 metres near New Guinea, whereas in the east off the South American coast, it can be as shallow as a few metres and may even reach the surface in places.

This broadbrush picture of the Pacific can be considered the 'normal' state of affairs. However, it would be misleading to think of the Pacific as stable. In the tropics, the surface of the ocean and the atmosphere directly above it are in constant dialogue, with the state of one constantly influencing the state of the other. The winds change the sea-surface temperatures, which in turn change the pattern of rainfall and winds. But the messages passed between the two mediums are not communicated at the same speed. While the oceans contain enormous heat capacity, they also possess huge inertia. The ocean's reflexes are slow, and are consequently in a perpetual chase to catch up with the atmosphere. The tropical Pacific, like all tropical oceans, is therefore never quite in equilibrium. At times, the equilibrium may become extra sensitive to disturbance. This is most likely to occur around March and April when the trade winds are at their weakest. It may also follow an intensification of the usual east-west differences in surface pressure and temperature. At such times, even the gentlest of tickles to the Pacific's sensitive nose might be enough to generate the most violent of sneezes.

No one knows for sure what initiates an El Niño event. At various times, the finger has been pointed at changes in the

Indian Ocean, the snow cover in Asia, and even the Antarctic pack ice. Other theories have suggested seismic activity as the trigger, linking the apparent correlation between particular eruptions and the onset of El Niño events. While these suspected triggers far from the tropical Pacific may have the potential to modify El Niños, most scientists now believe the real culprits lurk much closer to home in almost total anonymity.

For example, the trigger could be a pulse of unusual sea-surface temperatures. It may be the passing of a sub-surface wave, still reverberating from the last El Niño event. It might also be a sudden burst of the westerlies that interrupt the trade winds from time to time, before disappearing just as mysteriously. It was just such a meteorological reversal in the western Pacific that fired the starting gun for the 1997–8 El Niño. And the same phenomenon miraculously pulled Sir Francis Drake off a Sulawesi reef, where the trade winds had skewered his ship in January 1580, threatening to drown his crew and cargo of spices.[3]

The nature of the trigger is not the most important factor. It may vary with each new event or occur in a different location. Likewise the more usual order may be reversed with a La Niña preceding an El Niño. In fact, any number of minor anomalies have the potential to set off a chain reaction.

Whatever the trigger, a self-perpetuating loop between the ocean and atmosphere is initiated. In the case of an anomalous burst of westerly winds, the winds prompt a switch in the local current, which moves the warm water pool a little to the east. This causes a further relaxation of the trade winds that allows even more warm water to shift east again. The angle of the thermocline flattens, and less cold water is upwelled in the east. This decreases the difference in sea temperatures between east and west and similarly reduces the pressure gradient. Like a sumo wrestler caught off balance and pushed to the edge of the *doyo*, the interplay between atmosphere and the ocean can force the warm pool all the way across the equator to the boundary of the eastern Pacific.

At the peak of an El Niño, the broad picture of the Pacific is very different from the norm. The thermocline has flattened out considerably. So the deep, cold water normally close to the surface off South America is up to 30 metres deeper than usual. The sea-levels on both sides of the ocean are comparable. And the pressure difference between east and west has disappeared and, at times, even reversed. This is the flipside of the atmospheric seesaw that Gilbert Walker first noticed when he coined it the Southern Oscillation. Because the pressure difference drives the winds, the trade winds disappear and are replaced by westerlies that can blow nearly all the way to the Americas. The zone of cloudiness and heavy rain that normally characterises the western Pacific is now located across much of the equatorial Pacific, and in its place across Australasia the abnormally high pressures bring warmer, drier conditions. The clear skies associated with the usual zone of high pressure off the western coast of South America are replaced by rainclouds, as the warmer water off the coast heats the air above it.

The changes at the ocean surface are reinforced by two other very important and related processes evolving tens of metres below. The first is the excitation of a series of massive 'gravity' or Kelvin waves, which travel eastwards close to the equator several tens of metres below the surface. Their effect is to send 'packets' of warm water towards the South American coast, further depressing the thermocline in the east. The second, which is sparked in response to the Kelvin waves, is the generation of a series of another type of sub-surface wave, known as a Rossby wave. These have the opposite effect of Kelvin waves, heading westward and pulling the thermocline closer to the surface.

Kelvin waves travel around 100 km/day, and depending upon where they originate take up to two to three months to reach the South American coast. Once there, they are partially reflected up and down the coast – contributing enormously to the distribution of heat into the mid-latitudes – and partially reflected back across the Pacific as Rossby waves. The Rossby waves travel at about a third of the speed, taking between six and twelve months to reach the western boundary of the Pacific. From here they are reflected eastwards as Kelvin waves, this time as upwelling waves – the opposite amplitude from the initial series. But the coded instructions for an El Niño's demise may be hidden within the Rossby waves. According to the most widely accepted theory on the mechanism of an El Niño's decay, these reflected Kelvin waves provide the negative feedback necessary to end an event. In fact it is thought that the length of time it takes for these undersea waves to travel the breadth of the Pacific basin and back is the primary determinant of the roughly year-long duration of a typical El Niño event.

The signal to stop the physical processes that maintain the El Niño phase begins a new feedback loop, this time in the opposite direction. Easterly trade winds strengthen and warmer water is pushed further west, once more enhancing the pressure differential from east to west which drives stronger trade winds. The now-depleted warm pool in the western Pacific begins to accumulate heat again. After most but by no means all El Niños, this feedback system 'overshoots', instituting the La Niña phase of the overall cycle. When this happens, all the typical indicators of a normal Pacific are exaggerated. Trade winds are strengthened, the pronounced low pressure over Australasia brings heavier rainfall to the region, and the steeper thermocline means an even more vigorous upwelling from the Peruvian depths.

At its peak, the warm pool can stretch as far as one-third of the way around the globe, over a surface area one and half times the size of the United States. The extra evaporation, and hence extra latent energy, caused by the warm water being distributed over a much larger area than usual can raise the global temperature by over 0.3°C. In fact, the El Niño phases of the overall oscillation are commonly referred to by scientists as 'warm' events, and the La Niña phases as 'cold' events. One theory even suggests that an El Niño event serves as a release valve for the heat build-up in the warm pool, while a La Niña event acts to recharge it. According to one calculation, the loss of heat content from the equatorial Pacific during the 1997–8 El Niño was equivalent to the work of 900,000 twenty-megaton bombs.[4]

Such a profound shift in the earth's heat distribution is enough to alter the rotation of the earth. The relaxation or reversal of the trade winds causes less drag on the atmosphere from the mountains and the earth surface, allowing the atmosphere to spin round faster in the direction of the earth's rotation. In order to conserve its angular momentum, the earth compensates by slowing down by up to 0.7 milliseconds a day.

In view of this vast geographic shift in one of the earth's main climatic engines, and the interconnections between all atmospheric and oceanic cycles, it is understandable that reverberations can rattle around the entire globe. But the pattern of climatic impacts, or teleconnections as they are often called, that are generated by each El Niño can vary enormously between events. There are many factors that influence and modify the impacts on a particular region. These include the degree of sea-surface temperature change in the Pacific, the background climatic conditions, the timing of El Niño changes relative to the annual cycle and local and regional meteorological conditions.

Providing a general description of typical El Niño impacts is fraught with dangers and requires some qualification. Firstly, the presence of El Niño activity in the Pacific only affects the probability of certain meteorological conditions, and in no way guarantees they will occur. Secondly, reference to a particular impact on a region's weather should not imply that such impacts continue throughout the full length of El Niño conditions in the Pacific. Discernible impacts may be limited to only one part of the seasonal cycle, and are often only significant in influencing a region's rainy season. Thirdly, caution needs to be applied when attributing the causal effect of abnormal weather to El Niño, particularly in the mid- to higher latitudes. After all, severe climatic behaviour is a natural phenomenon and the occurrence of extreme weather even during an El Niño event may be more coincidence than consequence. However, there are many regions that receive a relatively consistent signal, where rainfall variability is amplified between a third and a half more than in areas of a weaker or inconsistent signal.

Because it is an essentially tropical phenomenon, most of the pronounced and consistent responses to an El Niño event can be found within those areas of the ocean basins that girdle the equator. In fact, throughout much of the tropics, El Niño accounts for much of the year-to-year variation in rainfall. Many of the anomalies in the Pacific region can be attributed to the direct effect of sea-surface temperature changes, whereas in the Indian and Atlantic regions, they result from changes in the tropical atmospheric circulation. As a consequence, the major convective zones shift from their usual locations over equatorial Australasia, Africa and Amazonia, to the equatorial central Pacific, central-western Indian Ocean, the Atlantic coast of Africa and north-western South America.

The effects in the mid- and high latitudes, on the other hand, are generally much more variable. The nature of the signal from the tropical Pacific to the higher latitudes alters from event to event, in response to variables such as the exact position of the newly located convergence zone and the timing of physical processes relative to the annual cycle. And although El Niño can account for a substantial proportion of the rainfall variability in the higher latitudes, the El Niño signal may be overwhelmed by other natural sources of variability.[5]

One of the important mechanisms by which the massive changes in the tropical Pacific are transmitted to the higher latitudes is through the effect such changes have on the jet streams. These are bands of very high-speed westerly winds that encircle the globe in the upper atmosphere of both hemispheres. Jet streams are very important to the weather of the higher latitudes because of their influence on the location of the dominant high and low pressure systems and their steering of storm tracks. The southern branch of the North Pacific jet stream, for example, is invigorated and pulled south during El Niños, thereby guiding winter storms on to the North American coast around California – much further south than their usual landfall on the Canadian border.

Typically during El Niños, rainfall is greatly reduced over much of Indonesia; the Philippines; northern and eastern Australia; the populated Pacific Islands; New Zealand's North Island; India; southern Africa; the Ethiopian highlands; Ghana; Nigeria; Sahelian Africa; most of central America extending into central Mexico, Colombia and northern South America; the Caribbean; north-east Brazil; and the Altiplano of Peru and Bolivia. Most of the major droughts in these regions have occurred during El Niños. The Sahel receives a moderately strong El Niño signal but is also markedly affected by longer-term fluctuations. On the other hand, heavier rain and increased probability of flood conditions are found over much of southern USA and the Great Basin; northern Mexico;

the Peruvian and Ecuadorian coastline; central Chile; south-eastern and northern Argentina; Paraguay; Uruguay and southern Brazil; the islands in the central and eastern equatorial Pacific such as the Galápagos and Nauru; New Zealand's South Island; the very southern tip of India and Sri Lanka; central China to southern Japan; Vietnam; the western cape of South Africa; Kenya; Tanzania; Uganda; and much of western Europe. There are many other regions, such as central California and northern Europe, that experience significant impacts, although the nature of the anomalies is somewhat inconsistent.

The consequences of rainfall fluctuations have been noted in the correlation between El Niño events and several natural phenomena. These include the flow of many of the world's great rivers such as the Nile, Amazon, Senegal, Parana, Yangtze, Orange, Krishna, Murray-Darling, and the levels of a number of lakes, including Lake Eyre, Great Salt Lake, Lake Chad and the Laurentian Great Lakes. The frequency of forest fires in Florida, Indonesia, south-eastern Australia and western Europe also correlates with El Niño events.

Temperature fluctuations do not generally impact on human societies to the same degree as variations in rainfall, but nevertheless they can have significant effects. Warmer than average conditions during El Niños are found over parts of northern South America, the Caribbean, north-eastern USA, eastern and north-western Canada, southern Alaska, Japan, south-eastern Australia, India, South-East Asia, southern Africa and Madagascar. Lower than normal temperatures chiefly occur in the USA's Gulf states and central south-western Pacific. In highland New Guinea, the lower temperatures associated with El Niños tend to bring heavier frosts.

In the USA, broadscale changes in temperature patterns also have an important influence on tornado activity. The southward movement of the sub-tropical jet stream, associated with El Niños, reduces the temperature differences between the Gulf coast and the Great Plains, thereby reducing the likelihood of tornado creation. As a result, El Niño events tend to coincide with weaker tornados with shorter damage paths than normal.

Broadly speaking, the pattern for La Niñas is the reverse of that of El Niños. Excessively wet or flood conditions are experienced predominantly over the continents bordering the Pacific and Indian Oceans – in particular, Australasia, northern China, India, southern Africa and parts of north-eastern South America. Dry or drought conditions tend to occur most commonly in the Gulf states of the USA, south-eastern Argentina, central Chile, central China, South Africa's western Cape, eastern Africa and much of western Europe.

Unusually colder temperatures are commonly experienced across northern South America, the Caribbean, Alaska and north-western Canada, Japan, South-East Asia, India, southern Africa, Sahelian and north-western Africa, and western Europe. North-eastern Australia and central south-western Pacific tend to experience warmer conditions.

The polewards shift of the sub-tropical jet streams in La Niñas increases the probability of stronger and more frequent tornado activity in the USA in all tornado-prone areas – particularly in the Ohio and Tennessee River valleys – with the exception of the Florida peninsula.

One of the most dramatic impacts of La Niña and El Niño events on people is the reshuffling of likely tropical storm formation zones. However, it is worth noting that while El Niños and La Niñas affect the probability of storms forming, it is impossible to blame them directly for the occurrence of any specific storm.

During La Niñas, high pressure cells in the sub-tropical Atlantic tend to be weakened and displaced off North and

South America, creating warmer waters than usual. This has the effect of extending the hurricane season. The frequency of hurricanes occurring during La Niñas is approximately double that of normal years. The coincidence of major hurricane years and La Niñas extends back as far as La Niña years can be dated. The year 1780 for example, known in historical sources as 'The Great Hurricane Year', has been listed as a La Niña.[6] During that year, eight major tropical storms hit the Caribbean, one of them being the deadliest Atlantic hurricane ever, reaping a death toll of over 25,000 people on Jamaica, Martinique and St Eustatius.[7] By the end of the 1780 hurricane season, the British, Spanish and French naval forces that were positioned in the Caribbean as part of a power struggle in the region had all suffered serious damage, setting back their respective ambitions.

La Niñas also increase the probabilities of hurricanes tracking beyond their normal range, with many extending further south into central America, and further north on to American landfall. Four of the last five major hurricanes to hit Nicaragua have occurred during La Niñas. Likewise, twenty-two of the sixty-six major hurricanes that struck the USA last century have coincided with La Niñas, compared with just five during El Niños. Furthermore, on one particular day during the 1998 hurricane season, by which time La Niña conditions were operating in the Pacific, no less than four separate hurricanes raged simultaneously. The clash of Hurricanes Karl, Ivan, Jeanne and Georges marked a record for the century, representing the highest number of hurricanes to occur at one time in the Atlantic. The quartet was followed only three weeks later by the arrival of Hurricane Mitch, the Atlantic's deadliest storm of the twentieth century.

El Niños on the other hand have the opposite effect on Atlantic hurricanes. The associated increase in cooler waters, as well as the southward movement of the jet stream that acts to shear off the tops of developing storms, reduces the average frequency of Atlantic hurricanes by half. Even so, the existence of El Niño conditions in no way precludes the formation of powerful hurricanes, as illustrated by 1992's Hurricane Andrew, in financial terms the costliest hurricane ever.

El Niños also affect tropical storm activity in the Pacific. In the north-east, the frequency of tropical storms does not necessarily change but they are more likely to originate west of 120°W. For instance, it has been calculated that the likelihood of tropical storms hitting Hawaii is three times greater in El Niños than La Niñas. The corollary is that during La Niña years, hurricanes are more likely to originate and remain closer to the Mexican coast.

In the Pacific south-west, the tropical cyclone tracks tend to move north-east of their usual location during El Niños, with fewer cyclones off the Queensland coast, while island groups to the east of about 165°E, such as French Polynesia and Samoa, are much more likely to be hit by cyclones. In the north-west Pacific the probability of typhoons occurring west of 160°E is suppressed during El Niños. In La Niñas the tendency is the opposite and it has been calculated there is a 20 per cent greater likelihood of typhoon landfall in southern Japan and the northern Philippines.

An examination of a single event reveals a unique collection of disasters and impacts. Two recent events – the El Niño of 1997–8 and the La Niña of 1988–9 – were among the strongest of the century and provide excellent examples of some classic responses.

The El Niño of 1997 was widely predicted by scientists. What was not foreseen was that it would evolve so rapidly from its beginnings around March–April to rival the 1982–3 El Niño as the biggest event of the century. The first great natural disaster blamed on El Niño came in July when eastern

European countries experienced their worst flood for over a century, affecting Poland, Hungary, Romania, Slovakia, the Czech Republic and eastern Germany. The worst floods in three decades hit Burma where half a million people were left temporarily homeless. In Africa, flooding in Kenya, Tanzania, Uganda and southern Somalia brought cholera and an epidemic of Rift Valley fever. There were heavy rains in the Galápagos Islands, while off the coast, the influx of warm water caused extensive bleaching of coral reefs. Peru was soaked by months of heavy rain, wrecking thirty major bridges and over 375 miles of major roads, resulting in a £60 billion repair bill. Eventually the country was saturated to such an extent that vast new lakes were formed as the land could absorb no more. A huge 300 by 40 kilometre lake appeared in the Sechura desert, making it Peru's second largest and earning it the local name 'La Niña'.

Droughts gripped parts of Indonesia, New Guinea, Hawaii, central America, Venezuela, inland Colombia, the Caribbean, northern China, North Korea, and the watersheds of the Mekong and Indus Rivers. In Indonesia, north-eastern Brazil, eastern Russia and Florida exceptionally dry conditions resulted in extensive fires. However, Australia, India and South Africa, three regions where severe drought was expected during the El Niño event, all received either normal or mild drought years as a result of the modifying influence of abnormally warm waters in the Indian Ocean.

In North America, the northern branch of the jet stream confined cold air to central Canada and allowed record warm temperatures across the Great Basin and northern plains, creating a £3 billion saving in winter fuel costs. Meanwhile a strong southern branch of the jet stream brought Pacific storms to the California coast and cool wet weather to most southern states. In January, an exceptional storm system extended between the lower Mississippi Valley through to south-eastern Canada. Floods and several tornados resulted in the south-east, while the north-east experienced the worst ice storm in over a hundred years. Some 80,000 miles of power and phone cables were brought down under the sheer weight of ice, blacking out over three and a half million homes. The damage bill came to £2.8 billion.

In the Atlantic, the hurricane season was very mild, featuring only three hurricanes compared to the average six, with a record of just one hurricane occurring during August–September, normally the months of peak tropical storm activity. On the other side of the USA, there were seven hurricanes compared to the average five, although two major hurricanes struck the west coast of North America.

As a general rule, the strongest La Niña events do not tend to clock up the same list of disasters as the strongest El Niños. This is partly due to physical reasons. There is for example no known evidence in the instrumental records of any La Niña event rivalling the 1997–8 El Niño in terms of sea-surface temperature change. But it is also a result of demographics. To many communities in the tropics and sub-tropics, La Niñas mean heavier rains and often a welcome extended monsoon season. The last strong La Niña of 1988–9 provides a useful profile of typical impacts.

The evolution of La Niña conditions began early in 1988, immediately after the demise of the 1986–7 El Niño. By April, northward displacement of the North Pacific jet stream resulted in the similar displacement of storm tracks into Canada, thereby denying the USA the storms that bring much of its rainfall. By July, 43 per cent of contiguous USA was categorised as suffering from severe or extreme drought, recalling the years of the Great Dust Bowl. Grain growers lost more than £19 billion, and the world's grain reserves were halved. Elsewhere, drought also prevailed in central and north-eastern Argentina, southern Brazil, the North China

Plain, Tunisia and Algeria. Western and much of eastern Europe meanwhile enjoyed the mildest winter for a quarter of a century.

For many more countries, 1988 is remembered as the year of great rains. Floods struck El Salvador, Nicaragua, Colombia, Pakistan, southern Thailand, South Africa, Mozambique, Nigeria and central China. In Australia, flooding followed a little later in early 1989, partially filling Lake Eyre. In northeast India, a flood on the Brahmaputra first destroyed parts of the state of Assam before submerging more than three-quarters of Bangladesh. The western Amazon recorded its worst flooding for two decades. In Japan, cool and excessively wet weather in summer resulted in the poorest crops for five years. In Sudan, the mean annual rainfall fell within fifteen hours during September, inundating over three-quarters of Khartoum and destroying over 70 per cent of the country's houses and 80 per cent of its schools. Fortunately for the rest of the Sahel, the year's rain was distributed much more evenly, with a wet summer contributing to the best year in two decades.

During a very heavy Atlantic hurricane season, Hurricane Gilbert set a record low in sea-level pressure of 888 millibars before going on to wreak havoc in Jamaica, Haiti and eastern Mexico. Hurricane Joan took an unusually southern storm track across the southern Caribbean before reeking a billion dollars' worth of damage in Nicaragua. But the La Niña's contribution to possibly the greatest headline-grabbing disaster of the year – the largest oil spill in US history – is virtually unknown.

When Captain Joseph Hazelwood steamed south out of Valdez Narrows aboard the Exxon *Valdez* on the night of 23 March 1989, his path down the Prince William Sound shipping channel was blocked by something highly unusual for the location – a wall of icebergs.

Just to the west of the Valdez Narrows lies the Colombia

Glacier, one of twenty glaciers that intermittently shed icebergs and growlers into the Sound. But despite the proximity
of the Colombia Glacier to the Narrows, its icebergs are
seldom expected to flow into the Narrows, or more importantly the shipping channel that bisects it. Iceberg movement
is controlled by the circulation of fresh water within the
Sound. Despite the great volume of melt water that flows
directly into the Sound, the main source of fresh water comes
from streams that directly enter the Pacific and are channelled
northwards up the coast by the extremely powerful Alaskan
Coastal Current. On reaching Prince William Sound, the
Alaskan Coastal Current splits and one branch enters the
Sound through Hinchinbrook Entrance on its south-eastern
side, circulates anticlockwise around the Sound, and exits
through Montague Strait in the south-west. In the process,
icebergs are normally dragged to the west away from the
Valdez Narrows and its shipping lane.

In March 1989, certain conditions in the Sound were
exceptional. The predominant controller of weather over
the Gulf of Alaska is the Aleutian Low, a permanent low-
pressure system that has a major influence on the strength of
the Alaskan Coastal Current. The behaviour of the Aleutian
Low is known to be strongly influenced by La Niña conditions in the tropical Pacific. Its location normally shifts from
east of the Aleutian Islands in wintertime to north of the
Bering Sea in summer. In La Niña years, the jet stream tends
to be weaker across the central Pacific, effectively weakening
the Aleutian Low and blocking its normal seasonal movement. For the first time since 1971, the Aleutian Low did not
move south of the Bering Sea during the La Niña winter of
1988–9. And in Prince William Sound, the current in March
1989 was half a knot, the slackest in nearly sixty years of
records.

Captain Hazelwood admitted later he had been drinking in

the immediate hours before sailing, a fact that left him shouldering much of the blame. But his admission is only part of the story. Man-made disasters are rarely the result of a single mistake. It was never proven that Hazelwood was actually drunk on the bridge, nor that his decisions and actions in the events leading up to the accident were erratic or unorthodox. Exxon's decision to downsize crew numbers, a lack of radar monitoring by the coast guard, and human error on the bridge in executing Hazelwood's instructions all contributed to the accident. And so did another little-questioned factor – the presence of heavy icebergs in a shipping lane supposedly free from such hazards.

Faced with a wall of ice, Hazelwood decided to divert around it by heading due south out of the shipping lane, then tacking back before reaching Bligh Reef. It was a perfectly valid decision. The ship immediately before his, along with many others, had taken the same route. But the Exxon *Valdez,* unlike the others, never made it back to the shipping lane.

3

The Art of Survival

Oh Beloved, where are you hiding
I wait for you,
What is this season without you
    Hindu poem to the monsoon

In 1892, the Australian writer Henry Lawson was sent into outback New South Wales. In a bid to separate Lawson from his city drinking buddies, his patriotic editor at *The Bulletin* assigned him to the country town of Bourke to write about the emergence of a distinctive 'bush' mentality. In his classic poem 'Bourke', Lawson described the conditions he encountered:

. . . No sign that green grass ever grew in scrubs that blazed beneath the sun;
    The plains were dust in Ninety-two and hard as bricks in Ninety-one.
    On glaring iron-roofs of Bourke, the scorching, blinding sand-storms blew.
    No hint of beauty lingered there in Ninety-one and Ninety-two.[1]

It is no coincidence that the years 1891 and 1892 were also El Niño years, and among the century's strongest. As we shall see later, extreme weather was being experienced simultaneously in parts of Russia, Africa and South America. And just as the effects of the drought proved a formative experience for one of Australia's greatest writers, climatic abnormalities the same year had a strong influence on Russian writers Leo Tolstoy and Anton Chekhov.[2]

Meanwhile, the poet Banjo Paterson, Lawson's friendly rival and colleague on the *The Bulletin*, was interpreting the same scenes from the outback with a much less pessimistic hue. Unlike Lawson, Paterson retained very happy memories of growing up in the bush, and had spent much time observing its 'many moods and changes'. In 1892, Paterson wrote:

So you're back from up the country, Mister Lawson, where you went,
    And you're cursing all the business in a bitter discontent;
    Well, we grieve to disappoint you, and it makes us sad to hear
    That it wasn't cool and shady – and there wasn't plenty beer . . .

    Yet, perchance, if you should journey down the very track you went
    In a month or two at furthest you would wonder what it meant,
    Where the sunbaked earth was gasping like a creature in its pain
    You would find the grasses waving like a field of summer grain,
    And the miles of thirsty gutters blocked with sand and choked with mud,
    You would find them mighty rivers with a turbid, sweeping flood . . .[3]

The period when Lawson and Paterson were writing was a rich one for Australian arts and literature. In the final decades leading up to federation, the visions of its white inhabitants had begun to turn away from a European past and look inland towards the soul of their new country. What they saw was an entirely unfamiliar environment, with a climate of a distinctively alien rhythm. Here, all the old relative certainties disappeared, and the climate could not be trusted. The timing of the seasons and reliance on the rains could no longer be taken for granted. Over time it dawned on the settlers that what this new climate provided generously in one year, it could take away in another.

Without knowing it, Lawson, Paterson and others had begun to describe the nature of El Niño. Awareness was blossoming that this was a land of unpredictability and extremes. And many of these extremes came with El Niño cycles. In a country where over three-quarters of all major droughts have occurred during El Niño events, the phenomenon has imparted a recurrent, hidden, profound influence on Australia's colonial history and culture, just as it has for many other nations.

Nothing prepared the colonists for the vagaries of the Australian climate. Despite aboriginal tales of great floods and droughts, the settlers were locked into their memories of Europe's predictable seasons, where at the most a few weeks of bad weather were always followed by a return to the regular pattern of previous years. The new land pulsed to a very different rhythm. There could be years of reasonable rains and moderate temperatures, enough for settlers to dig in and stake their future. There would even be the odd year of generous rain, when the grasses and crops flourished, and the settlers could cash in. Yet, another adverse year was never too far away, and the dreams and savings of many settlers would be ruined. Many tales from these early days highlight the cost to settlers of failing to appreciate the potential extremes of their new land.

For example, when the New South Wales town of Wagga was hit by the flood of the Murrumbidgee in February 1844, only two of its ninety-one inhabitants survived. One of them was a young girl strapped by her lover to a tree, and the other the publican away on business. The local aboriginal tribe, aware of the river's potential, all survived by climbing a hill in time. A replacement town was later built on higher ground on the other side of the river, where it now rejoices in the splendid name of Wagga Wagga.[4]

How the early colonists saw the inner heart of their continent and its potential was for decades determined by the heroic explorers that pushed further and further inland. The impressions they sent back fluctuated wildly, sometimes deceived by the heavy rains of a La Niña year, at others frightened by the spectre of an El Niño drought. Few seemed to grasp the complete picture in all its different hues. John Oxley's inland expedition of 1817, conducted immediately following flooding rains of the first few months of that year, turned back soon after reaching the Lachlan River, convinced the interior of the country was covered by an 'inland sea'. Towards the end of the 1835–6 El Niño, Major Thomas Mitchell found the same river 'down which my predecessor's [Oxley's] boats had floated . . . gone, with the exception of a few small ponds'.[5] Unfortunately, by then his team had dragged a set of heavy whaleboats 600 miles across land. Oxley's vision of an inland sea was seemingly confirmed on his second expedition, this time down the Macquarie River during the heavy autumn rains of 1818, concluding that all inland waters 'have but one common reservoir'.[6] Yet the subsequent expedition on the same river by Charles Sturt, in the El Niño drought of 1828, found 'hollow after hollow that had successively dried up', Sturt concluding that the region was 'unlikely to become the haunt of civilised man'.[7]

Whenever explorers sent back favourable reports, pioneering settlers were ever ready to follow in their footsteps. One of John McKinlay's party, for instance, during his expedition in early 1862 to search for the missing explorers Burke and Wills, found knee-deep grasses and the land 'being clothed in many places with a nice green coating'.[8] The description lured many unsuspecting newcomers into the region that is now known as the Great Stony Desert. Over the decades the irregular cycle of intermittent good and bad years saw tides of pioneering farmers and pastoralists advance and retreat across Australia's arid margins. It would only take one good year before the last drought was quickly forgotten, the land again overstocked, and the stage set again for further misery and loss come the next El Niño. One of the characters in George Johnston's *My Brother Jack* thus described the continent's inner heart: 'Here journeys have ended, the pioneering flame has guttered and failed, hopes and ambitions lie buried beneath the blowing sand.'[9]

This ignorance of the climatic cycle can be blamed partly on the lack of long-term historical data, but it was also partly a refusal to acknowledge the worst in a misplaced patriotism towards the new land. The convict past was being left behind, and the growing sense of nationalism coloured many a settler's image of the country with an optimistic light. Droughts, floods, rabbit plagues and other natural disasters were seen, particularly by the authorities, as unpatriotic. Even as late as 1919, a film of New South Wales's drought-ravaged plains during that El Niño year was banned from overseas distribution by state authorities for unpatriotically exposing the 'disastrous position'[10] of the state. Few wanted to see drought and flood years as anything more than an annoying aberration, and not as an integral characteristic of the climatic cycle. 'Rain follows the plough' was the prevailing thinking, and as a consequence droughts could be overcome and eradicated. There was a belief that man could shape the land into the character of a new Britain, but, imperceptibly, it was the character of many of the British colonists that was gradually being moulded by the land.

While Lawson, Paterson and others were among the vanguard of artists to depict the distinctive climatic rhythm to the land, they were also among the first to encapsulate and celebrate the emergence of a distinctive Australian 'character' to the bushfolk. As another verse from Lawson's 'Bourke' of the time of the 1891–2 El Niño suggests, Lawson and Paterson's two favourite themes were related.

> Save grit and pulse of generous hearts – great hearts that broke
> and healed again –
>   The hottest drought that ever blazed could never parch the
> souls of men;
>   And they were men in spite of all, and they were straight,
> and they were true;
>   The hat went round at trouble's call in Ninety-one and
> Ninety-two . . .[11]

This influential vision of an Australian 'character', later adopted by city-livers eagerly searching for a new identity, was of a laconic, stoic, heroic, gambling pragmatist. Above all, though, he was a 'mate'. No value was placed higher than that of dependability in a crisis. Everyone needed 'mates' to rely on in the searingly harsh, isolated beyond, because bad times were always just around the corner.

That such an identifiable character emerged not from the urban majority but from those that bore the brunt of El Niño's extremes was no coincidence. For many in a modern industrialised world, it might be difficult to appreciate the potential devastation that can be wrought by a climatically adverse year. Diverse sources of food and revenue, as well as extensive trade and transport networks, are part of a vast array of mechanisms that shield urbanised societies against much of the shock of climatic fluctuations. But for people living directly off the land, natural disasters and environmental extremes have always been a major source of risk. Their culture is part of an integrated system in which the population has adapted to the environmental risks, and has prescribed measures to ensure long-term survival.

How much of those cultural adaptations to climatic variation can reasonably be ascribed to El Niño is a very difficult question to answer. Strategies have evolved to cope with the risk of a fluctuating food source, whatever the ultimate source of that risk. In addition, episodic shortages result from many other non-El Niño sources of climatic variability such as longer-term cycles or random variation, as well as non-climatic sources such as earthquakes and epidemics. But there are many cultural adaptations throughout the world that can clearly be seen to have evolved primarily in response to year-to-year variability. A closer look at the nature of these adaptations and the geographical locations in which they have evolved clearly points to El Niño's influence as a major selective pressure.

Regions under a strong El Niño signal, which generally speaking are located across the tropics as well as around the basins of the Pacific and Indian Oceans, experience high interannual variability as a dominant feature of their weather. It has been estimated that, compared to regions that do not receive a strong El Niño signal, annual rainfall variability is amplified by between one-third and one-half. The reality for people living in such regions is more frequent episodes of crop failure and deprivation. Even more significantly, before El Niño predictions were available, people had no idea when the extremes would occur. The only certainty was that on average one in every few years would be poor, one in every ten to fifteen years would be particularly severe, and once or twice a generation, nothing short of catastrophic. Those that chose to deviate from the stringent cultural constraints imposed on them by the variable environment may have enjoyed the fruits for many years, but come the catastrophic year, migration or death would have been their only options.

Historically, the societies most beholden to climatic fluctuations were traditional hunter-gatherers. With limited ability to manipulate their environment, these societies existed at the mercy of wild swings in food availability over space and time. A glimpse at tribes that have survived into modern times, such as the Australian Aborigines and the !Kung of the Kalahari, reveals many social characteristics that are common to all of them and that have evolved primarily to cope with an extremely variable and unpredictable rainfall.

A perception persisted throughout the first half of the twentieth century that these societies were made up of savages desperately eking out a meagre existence on a marginal food supply. By the time serious anthropological studies were undertaken, particularly in the 1950s and 1960s, impressions of a constant struggle to stave off starvation were hastily laid to rest. In contrast, life at the very margins of survival was shown to be surprisingly comfortable. Typically, only two to five hours per day were needed in food procurement, even in the harshest environments, and in some cases more than 40 per cent of the group were not even involved. As a consequence, there was always an abundance of leisure time, and nutritional status was excellent. The new picture that emerged of hunter-gatherers was well encapsulated by the term commonly used by anthropologists, the 'affluent society'.

Now, in the light of more recent long-term studies, a more complex and less romantic picture has emerged. It has become apparent that the average life of a traditional hunter-gatherer could more accurately be described as generally undemanding, yet punctuated with irregular periods of severe stress. As primary determinants on social organisation and population density, these bad periods apply a constant brake.

Cultures that have adapted to maximise the returns of an average year risk the rare but inevitable occurrence of widespread famine and potential cultural collapse. In order to avoid these worst case scenarios, regulatory cultural mechanisms have evolved to ensure access to resources even in the bad years. These mechanisms take a variety of forms, most of which are adaptations to fluctuating resources in marginal environments.

Mobility or nomadism was an essential adaptation used to track the erratic supply of food and water over large distances. This restricted these hunter-gatherer societies in a number of ways. Possessions needed to be kept to a minimum, and 'tool-pouches' confined to a few, simple light essentials. Mobility also made large groups of people impractical so small bands averaging fifty individuals became the norm. An important premium was placed on the memory of coping with past disasters so that remedies could be found. Knowledge of substitute wild foods, drought refuges, hidden springs, long distance personal connections and appropriate cultural responses were coded into traditions and myths for future reference.

The periodic scarcity also dictated low population density, well within the 'normal' carrying capacity of the land. This prompted stringent constraints on reproduction through a combination of religious and sexual taboos: the use of natural contraceptives, limits on the frequency of intercourse and the extension of lactation. Even infanticide was practised.

Two opposing forces, each with severe drawbacks, helped determine population density. One was to exploit the resources available in the average years, leaving little to fall back on in times of drought and therefore exposing the people to starvation. The other was to maintain the population at a low enough density to survive even the most catastrophic years, leaving it vulnerable to attack and requiring excessively onerous cultural restrictions to keep numbers down. The solution, as anthropological evidence has shown, lay somewhere in between. This compromise allowed communities to cope with the years of moderate to average severity, but required a further strategy to weather the worst years.

These years made it essential to have secure rights to resources beyond a community's normal range. One option was of course warfare, but raiding was inherently risky, particularly for drought-weakened warriors, and could not guarantee success in the long term. The safer and historically preferred option has been the formation of special bonds, or alliances, between communities.

Alliances should not be confused with the strong sharing ethic and egalitarian spirit that characterise traditional hunter-gatherers and that act to buffer against natural daily and weekly fluctuations in food procurement. Fluctuations on a year-to-year scale imposed themselves simultaneously on the whole community and across whole regions, rendering the ability to share resources within the group of little use. Bad years created a premium on co-operation between communities, ideally from different ecological zones. These alliances acted as an insurance policy, pooling risk across a wide geographical range through the banking of social obligations to be drawn on in times of scarcity. Inter-community marriage ties formed the basis of many of these alliances, but there were also several other forms, determined by the particular ecological needs of the communities. But whatever the nature of the alliance, the common thread between all of them was the principle of reciprocity. It is possibly through these alliances, and their specific adaptations to inter-annual variability, that El Niño has generated its strongest cultural response.

The ethic of 'mateship' that developed in colonial Australia may be seen as an embryonic example of this reciprocity. Interestingly, in less than a century of exposure to the peculiarities of the Australian environment, the colonists were already acknowledging the emergence of 'mateship' as celebrated in some of their best- known literature. Contemporary ecologist and writer Tim Flannery has since recognised the adaptive benefits of 'mateship' in his much-acclaimed book, *The Future Eaters*. 'Perhaps there is something quite fundamental about such social obligations that makes them indispensable in the Australian environment.'[12] To the indigenous Australians, the system of reciprocity has developed over tens of thousands of years to become a central and inescapable part of traditional aboriginal life.

Ancient invisible ties bond all Aborigines to each other and to their land. An intricate web of ancestral myths, or Dreamings, serve as connections across vast distances. These were immortalised by Bruce Chatwin in his best-selling book, *The Songlines*. 'In theory, at least, the whole of Australia could be read as a musical score. There was hardly a rock or creek in the country that could not or had not been sung. One should perhaps visualise the Songlines as a spaghetti of *Iliads* or *Odysseys*, writhing this way and that, in which every "episode" was readable in terms of geology.'[13]

The myths, weaved in and out of each clan, are central to traditional aboriginal religion, and track the 'footprints' of revered ancestral creatures. Ownership of these Dreamings is a fundamental part of each individual, and through them each person interprets their relationship to the earth. Intermarriage also forges alliances between neighbouring clans, but the connection through totemic ancestors stretches even further to potential refuges great distances away.

The Aborigines' ability to bridge huge distances was vital in connecting people faced with drought to others unaffected by the same predicament. Without these distant bonds, those in need would have found their potential saviours in the midst of the same adversity. Dreaming tracks therefore stretched extraordinary distances, their extent and complexity reflecting the broadscale nature of the El Niño influence across the Australian continent.

These bonds and alliances would have been unnecessary for survival purposes for many years on end. Therefore an adhesive force was needed in the interim years to maintain them. Considerable time was spent in reinforcing the religious beliefs and social customs that held them together. Visiting distant relatives and neighbouring clans was essential, even in the absence of an immediate environmental necessity. This custom of going 'walkabout' for no obvious reason led white settlers to perceive their Aboriginal stockmen as unreliable and shy of hard work, a misconception that spread across the country. In times of great food surplus, such as explosions in edible rat populations and the abundance of seeds and fruits that follow the heavy rains characteristic of a La Niña year, contact with neighbours took the form of huge congregations, or corroborees.

The community's prospects for long-term survival hinged on adherence to its religion. Because the maintenance of tight bonds across large distances required considerable energy and costs, there were undoubtedly individuals that chose not to observe them. These nonconformists soon found themselves isolated from the rest of the community or subject to severe penalties such as denouncements, expulsion and even execution. In times of crisis, these individuals could expect no help from their kin.

Most droughts did not necessitate the immediate activation of these totemic relationships. These were reserved as last-gasp insurance. There was a hierarchy of strategies that could first be deployed. A comprehensive knowledge of game and drought foods, such as the roots of the desert kurrajong, enabled clans to travel large distances with surety. As drought took hold, clans could then retreat to the most permanent of the waterholes within their range. Some of their refuges were in sacred areas, where hunting and food gathering were prohibited. These acted as reservoirs for game that could help restock depleted areas. But once or twice in a generation, the community was forced to abandon its normal range entirely and draw on the alliances forged by thousands of years of survival in such unpredictable rainfall.

The basis of the social obligations maintained by the !Kung Bushmen of the Kalahari to protect against bad years is built not upon spiritual ties, but on *hxaro*, a reciprocal system of gift-giving. Over time, each !Kung member develops a strong bond of friendship with other non-related individuals through a delayed exchange of presents. At any one point in time, most of a !Kung's possessions will have been accumulated through their *hxaro* relationships. Gifts are received, kept for a while, sometimes repaired and often embellished, but then passed on again, even sometimes ending up back with the original donor. The only payment required is an unspoken sense of obligation on the part of the recipient to reciprocate at some point in the future. Once the bond is established and secure, each participant is said to *//hai* the other, literally meaning to 'hold the person to their heart', or in other words, to be always there for them in times of trouble. In a region subject to major and not infrequent droughts, such a system of drought insurance is absolutely essential for long-term survival.

The quantity or cost of the object of exchange is not directly important. Any non-food item might suffice, such as ostrich egg-shells, arrows, blankets, necklaces, pots or clothes. But that is not to say any item will do. Its value is calculated on its usefulness to the recipient, and much time is spent in discussing who has given what and to whom. Anthropologist Polly Wiessner, the first person to document the *hxaro* system, calculated that over 60 per cent of the conversation in a given month was focused solely on this topic. Stinginess or a tardiness to reciprocate is soon noted, and if ridicule, outrageous lies or other accepted forms of admonition do not produce results, then the partnership is discreetly dropped.

The choice of whom to have a *hxaro* relationship with is carefully considered in order to pool the risk across as wide a distribution of partners as possible. Visits to *hxaro* partners are frequent, and when they live far away, it is essential for the health of the relationship that contact is maintained through a balanced and continuous flow of gifts. Even spouses will maintain separate networks of *hxaro*, which can result in families being spread across hundreds of square kilometres in times of drought.

Wiessner witnessed the activation of the *hxaro* obligations for herself in the La Niña of 1974, when high winds ruined the crop of mongongo nuts and sustained heavy rains triggered both a bout of cattle disease and a destructive plague of insects. Game had dispersed away from the waterholes and, with nothing to hunt or gather, the people were forced to spend weeks of idle time sitting around talking about the strange weather and making handicrafts for *hxaro*. Within a few weeks, as the food shortage became acute, half of Wiessner's study group had scattered in all directions to visit their exchange partners because, in their own words, 'they missed them and wanted to do *hxaro* with them'.[14]

In Alaska's far north-west, the Nunamiut and Tareumiut tribes traditionally exhibited another form of alliance, one based on special trading relationships. In the case of these people, the vulnerability to climatic variation, and the need for interdependency between the two tribes, stemmed from their respective reliance upon a single food resource. The coastal Tareumiut life centred on bowhead whales, which migrated annually as they still do between the Bering Sea and the summer feeding and calving grounds of the Beaufort Sea. Each spring the migration route was confined to the leads of open water between ice-floes, taking the whales increasingly closer to shore until they were funnelled within range of the Tareumiut and other whaling communities. Further inland, in the forests of the Brook Ranges, the Nunamiut's traditional movements were based on tracking the seasonal migration of caribou.

Both the numbers of whales and caribou fluctuated wildly from year to year. This variation is linked to climatic changes to which El Niño is but one contributor. Although the Nunamiut's caribou numbers varied from year to year, the main fluctuations occurred in two cycles of a 15–20 year and 60–100 year frequency. The El Niño cycle probably showed a stronger correlation to fluctuations in the bowhead catch, not on whale numbers, but through its influence on Arctic pressure systems that would have in turn affected sea-ice floes and the accessibility of whales to the whalers. In certain years whaling was seriously impeded by constant storms and fogs that prohibited hunting, or off-shore winds that crushed the floating ice-pack against the shore, pushing the leads beyond easy reach of the whalers. In other years of more favourable off-shore winds, when the floating ice-pack was pushed back into the Arctic basin, leads occurred very close to the shore ice, allowing for exceptional harvests and great surpluses of whale meat to trade. If either the caribou or whale season were poor, the respective communities would have faced starvation had they not secured guaranteed access to a second food resource.

While intermarriage was a common enough practice between neighbouring communities, rarely did such kinship links extend beyond ecological zones. If a man took a wife from outside his own community, it was imperative she knew the necessary survival skills that were particular to his own environment. Across the territorial boundaries that separated the Tareumiut and Nunamiut, the most common alliance was with trading partners, or *nyuuviq*, as they referred to each other. There was no limit to the number of partnerships one could have, and ideally, financial considerations permitting, one would seek to establish a network of such links across as wide a geographical area as possible. The immediate benefit for the partners was that each could rely on the other for goods unobtainable within their own territory. The Tareumiut received such valuables as caribou skins for clothing, while the Nunamiut gained whale oil for fuel. Annual surpluses in any goods were reserved for one's *nyuuviq* and a spirit of generosity reigned over typical exchanges. *Nyuuviqs* could also be turned to in times of crisis, typically after the failure of the primary food resource, when survival meant refuge in another territory. And in such times, it was accepted practice to provide not only food and shelter, but the *nyuuviq*'s wife.

Another form of social insurance against wild year-to-year swings in food availability was the holding of great communal feasts. This practice was maintained by hunter-gatherer tribes along the north-west Pacific coast of Canada. Potlatches, as the feasts were known in the region, were held by village chiefs or the heads of several families to mark life-changes such as puberty, marriage and name-giving. Hosts invited other communities to witness these changes and participate in the lavish celebrations. The occasions served to highlight status and distribute an abundance of food, and what the guests could not consume they were encouraged to take home. There was always an unspoken obligation on the recipient to reciprocate the potlatch or lose enormous prestige. The feasts served to bank food surpluses into greater status with other communities, therefore providing an investment policy for leaner times.

In a region prone to considerable fluctuations in the traditional food sources such as salmon and berries, the potlatches may have functioned to distribute resources evenly across the communities.

The term used by anthropologists to describe hunter-gatherer groups such as those along the north-west Pacific and the Tareumiut and Nanumiut is 'complex', as opposed to the 'general' hunter-gatherers that are typified by the !Kung and the Aborigines of Australia. Complex hunter-gatherers are characterised by material ownership, economically-based competition and social hierarchies. They also tended to live in stabler, more productive, and often more seasonal, environments that encouraged the development of storage systems and allowed for less mobility and higher population densities. From their earliest beginnings, 20,000 years ago, developments in plant husbandry and the domestication of game saw agricultural societies develop out of 'complex' hunter-gatherer systems.

These technological innovations across the centuries achieved greater control over the natural world and provided some security against climatic variation. While the weather still determined the success of crops, plant husbandry was less affected by fluctuations in rainfall than traditional hunting and gathering had been. Storage also provided a safety net that could offset weeks or even months of food shortage. All these developments eroded the need for inter-community alliances and the tremendous investment required to maintain them.

For example, some of the //Gana hunter-gatherers living close to the !Kung in the Kalahari have been able to provide against short-term failure in the rains with the addition of some animal husbandry and food cultivation. By keeping a few goats and growing beans, maize and marotsi – a large water-bearing melon that will hold moisture for a few months into the dry season – they have adopted a more sedentary life style and reduced the time spent on maintaining inter-community relations.

But innovations in food storage and agriculture by no means removed the threat of year-to-year fluctuations. The inevitability of disastrous years at irregular intervals continues to overwhelm the 'insurance plans' of simple agricultural communities. As a result, those living in regions consistently prone to interannual variability still operate inter-community alliances.

Shortfalls in the rice harvest of Borneo's Dayaks are offset by the redistribution of grain between a large network of longhouses. Households that avoid the opportunity to share the good fortune of a rice surplus in ritual feasts are socially ostracised. They are viewed not merely as selfish, but as breaking the fundamental system of reciprocity between the human and spirit worlds that determines harvest success or failure.

In southern Africa, the Tswana people operate a system called *mafisa*. Bonds are formed by loaning cattle to others, who herd them and use their milk and draught power. Cattle are widely distributed through this arrangement, and in return for the loans owners are guaranteed shelter and hospitality in times of need.

Perhaps the best example of an alliance system evolved within an agricultural society to counter the shortfalls in harvest specifically associated with El Niño events is found in the highlands of New Guinea. Here, the unusual high-pressure systems that settle over the western Pacific during such events can bring not only drought, but frost. This rare phenomenon in the humid tropics can have dire implications for the primary crop of sweet potato. The sweet potato is planted in large mulch mounds, elevating the plants above the zone of lowest temperatures and providing a degree of frost insulation. However, in certain years, the frosts are so intense and persistent at altitudes above 2,250 metres that the mounds are totally inadequate. In such years, all surface growth is killed, and as a highlander succinctly explained in pidgin to a missionary during the 1972 frost: 'Everything is buggered up finish.'[15]

The highlanders have created a three-stage insurance plan determined by the frost's severity to cope with all contingencies. The first back-up, often useful after the minor frosts that can occur annually, is to maintain two gardens, one in the more fertile but frost-prone valley bottoms, and another on the slopes. The second is to acquire rights to gardens in adjacent valleys, usually within a day's walk. Although these second gardens are often at a similar altitude to the first, the frost damage can vary considerably between valleys.

The third plan, essential for the once-in-a-generation frosts such as those of 1941, 1972 and 1997, relies on kinship links that have been deliberately forged with tribes at lower altitudes, generally several days' walk away. The women and pigs are the first to move, often after four to five frosts in a row, with the menfolk following a few weeks later. Only the old and infirm stay behind. Once they have reached their new hosts, the refugees are automatically awarded rights to mature gardens, and may remain for up to several months or, if their former gardens cannot be quickly resurrected, even a few years. The welcome given to the highlanders by their lower-altitude hosts is by no means charity. It is based upon years of exchanges such as bride payments, and offers to share in the pandanus nut harvest, a highly prized crop in the highlands.

Those without kinship hosts can still migrate, but they must pay for their stay with pigs and can expect an abrupt end to the hospitality when the supply of pigs runs out. The need to migrate in times of severe frost was clearly demonstrated during the devastating El Niño of 1941. One clan in the central highland village of Yumbisa chose not to move because, just prior to the onset of frosts, they had been at war with their neighbours and feared being ambushed. When those that migrated eventually returned to the area, they found only two of the twenty-five-strong clan still alive.

With the profound effect that year-to-year climatic variability has had on various cultures, it is reasonable to wonder how much the climatic fluctuations of El Niño may have influenced the modes of food gathering that have been adopted for a given area. Could the degree of inter-annual variability have been a factor in determining where agriculture evolved? In a highly unpredictable environment, any shift towards a more sedentary existence would have resulted initially in less security in food supply. In addition, an unpredictable climate may not have suited the domestication of plants requiring regular seasonal rainfall. With more time needed to devote to food procurement, and less time available for maintaining alliance networks, it is probable that areas of the most uncertain yield would be the least likely in which agriculture would evolve.

Would hunter-gatherer groups enjoy an advantage in climatically variable areas compared to simple agriculturally based societies? The flexibility that comes with greater mobility, lower population densities, greater diversification of food supply, open social groupings and more extensive inter-community alliances suggests that they would. Of course, other factors would have also exercised a strong influence in the distribution of hunter-gatherer groups, such as isolation from agricultural neighbours, the lack of fertile soils, and the absence of plants and animals suitable for domestication.

It is perhaps significant that the vast majority of hunter-gatherer cultures that have survived into present times exist in zones of the strongest El Niño influence. The Bushmen of the Kalahari, the pygmy groups of equatorial Africa, the Hadza of East Africa, the Aborigines of Australia, the hill tribes of India, and various peoples from the Andamans, Philippines, Indonesia and Amazonia all live in regions in which the rainy season cannot be relied upon. The only exceptions are the Inuit people of the Arctic who do not subsist in a region with a consistent El Niño signal, although they too live in a highly variable environment. If El Niño has been a significant influence in the distribution of these cultures, then the more enlightened perception that views them not as 'primitive' relics but highly adapted societies gains added credence.

The emergence of civilisations has changed the way climatically abnormal years have affected people and societies. Risks that were once pooled between communities became underwritten by the state. Instead of investment in alliances through constant ritual, visiting, intermarriage, sharing or gift-giving, the new premium to insure against variation came in the form of taxes and tributes to the state. The principle of reciprocity, which was so strictly enforced by the keepers of inter-community alliances, vanished with civilisations because state leaders failed to keep their side of the bargain.

The vast majority of the peasants and the urban masses still had good reason to fear the consequences of climatic failure. The inflexible and exploitative systems of social and economic organisation created great inequalities among populations. As the economic historian Richard Tawney said of 1930s China, 'There are districts in which the position of the rural population is that of a man standing permanently up to the neck in water, so that even a ripple is sufficient to drown him.'[16] Traditional means of coping had been abandoned, along with ties with distant communities. Ordinary people lost the ability to survive on wild foods and game. In times of stress, the problem for many was not an inability to find or produce food, but an inability to acquire it. The rise of civilisations heralded a new era in vulnerability.

In theory, the technology and level of organisation essential to maintain a state led to certain powers over the environment. A variety of social and economic institutions cushioned the impact of minor climatic variations, so that fluctuations in weather no longer ruled daily lives. Yet as the mastery over nature improved, populations increased – often beyond the state's capacity to feed them. This vulnerability was most exposed in times of stress, particularly when climatic crisis coincided with other stresses such as epidemics, rising taxes, revolts, invasions and wars. In these times, harvest failure could prove the final straw, and the greater the population density, the greater the potential for catastrophic famine.

People sought a variety of explanations for the climatic extremes they most feared, often looking to the spirit world for answers. A host of religions and beliefs were adopted to help dissipate the anxiety, along with a multitude of customs. Each climatic disaster tested a community's confidence in its beliefs and rituals, rendering it susceptible to new alternatives. These disasters, many of which are now known to have resulted from the El Niño cycle, became synonymous with a battle for the souls of the masses.

> . . . they take the flow o' th' Nile
> By certain scales, i' th' pyramid; they know
> By th' height, the lowness, or the mean, if dearth
> Or foison follow. The higher Nilus swells,
> The more it promises. As it ebbs, the seedsman
> Upon the slime and ooze scatters his grain
> And shortly comes to harvest.
> *Antony and Cleopatra*, William Shakespeare[17]

The day the Nile began to rise was historically the most important in the Egyptian calendar. It was known as the Night of the Drop, when 'the angel on that day, throws a drop of water of such fermenting quality into the river, that it causes it to rise to such a height, as to overflow all the country'.[18] So began a time of great anxiety for the nation, as the people begged for *wafa*, or fulfilment of the promise of Allah. Each year, careful measurements were taken on the Nilometer, an elaborate, Escher-like four-sided stone well with steps down the sides spiralling around a calibrated column in the centre. Underground conduits allowed the water to enter from the Nile. If the level approached the height considered optimum for the prosperity of the people, the guardian of the Nilometer would swim to the column and anoint it with a concoction of saffron and musk. The next day, *wafa* was declared to the public through singers positioned around the Nilometer, and the guardian dressed in a robe of gold thread would lead a great procession through the crowded streets. A signal was given and the feeder canals would be opened to inundate the fields.

It is well established that the basis of Egyptian civilisation was rooted in the Nile's summer flood. The health and prosperity of the whole nation hinged on the floods reaching optimum level. A system of canals, levees and dams allowed the water and life-enriching silt to wash annually over the entire valley and irrigate the produce that supported the Pharaohs and their great kingdoms. According to the Egyptians, the Nile's annual rise and fall reflected the order of the universe. Before the source of the Nile was discovered in the mid-nineteenth century, not only was the origin of the river a complete mystery to the Egyptians, but so was the cause of its annual flood. They never knew that far to the south in the Ethiopian highlands, monsoon rains feed the headwaters of the Blue Nile and Atbara River each summer, supplying 95 per cent of the flood's volume.

The Nile is one of the world's most dependable rivers, usually recording only minor variations in its flood level. This very reliability was the crucial factor in the creation and success of the Egyptian civilisation. Yet every few years, the Egyptians' lifeblood would be switched off prematurely. These years of poor flood often correlate with El Niño's southward displacement of the rain-making systems from the Ethiopian highlands. Even small fluctuations in the flood level are enough to induce a catastrophic impact. During the 1877–8 El Niño, for example, the flood was only six feet below normal, yet in some provinces more than half of the valley was left dry. Historically, for Egyptians living along the valley – which even today supports over 96 per cent of the population – the failure of the floods and subsequent famine remained a constant anxiety.

> Hail to you, Hapi . . .
> When he is sluggish noses clog,
> Everyone is poor
> As the sacred loaves are pared
> A million perish among men . . .
> When he floods, earth rejoices
> Every belly jubilates
> Every jawbone takes on laughter
> Every tooth is bared
> 'Hymn to Hapi'[19]

The pendulous breasts of Hapi, the god of the inundation, were for many the ultimate sources of the floods. No one knew exactly where Hapi resided, but he belonged to the underworld, from where her waters were thought to flow out between two rocks in an underground cave. Each year she was entreated with flowers and prayers. Books containing spells were also tossed into the river and gifts cast adrift on elaborate shrines.

The search beyond the earthly world for the cause of climatic abnormalities was virtually universal throughout historic and prehistoric times. The smaller the community, the greater the tendency towards attributing the control of the weather to ancestors and living spirits. Great civilisations, on the other hand, understandably adopted more grandiose ideas, tending to link the vagaries of nature with deities that controlled the universe. The underlying purpose of these diverse beliefs lay in a human need to explain the natural forces that dominated life and preoccupied human fears.

Although people over the centuries have looked to the heavens and the afterlife as the ultimate controllers of weather, many cultures have attributed the behaviour of the living as the cause of adverse weather. To many Hindus, for example, monsoons are a gift of the gods, and their timely arrival and deliverance of an annual package of life-essential waters are considered a judgement on the pious and proper conduct of the people.

Nowhere has the connection between the weather and behaviour of the people been more integral to a nation's culture than in China. Even today many rural Chinese believe the moral actions of the people influence the behaviour of the weather. In the past, every individual's actions were thought to count, so that overwhelmingly moral behaviour by the majority could counteract the improper conduct of a few. Yet the actions of certain individuals held more sway than others, and none more so than those of the emperor.

The right of all China's emperors to rule was predicated upon the concept of *tienming* – the Mandate of Heaven. Each emperor ruled according to a divine authority to mediate on heaven's behalf, but only so long as he or she administered well. Heaven's warnings arrived in the form of great natural disasters, or *tianzai* as the Chinese called them, literally translating as 'heaven's calamities'. These warnings put the spotlight on the emperor's behaviour, and any sign of immorality, corruption, incompetence or indifference to the plight of his subjects brought his divine mandate into question. If warnings went unheeded by the means of repentance and remedial action, the right to rule could be transferred to a new dynasty.

The concept was first used by the Chou rulers after 1100 BC as justification for their overthrow of the Shang Dynasty. It found full voice in late Imperial times following the teachings of Confucius and Mencius who preached that a truly moral ruler was one who took all appropriate actions to provide for the well-being of his people. And even though the ideology has largely disappeared from modern Chinese thought, such recent disasters as the 1976 Tangshang earthquake were still interpreted by many rural folk as being directly connected to the death of Chairman Mao.

No country's civilisation has suffered more than China's. A close examination of its historical records reveals an incidence of drought, flood, earthquake, famine, epidemic or locust plague in virtually every year at least somewhere in the empire. But in human terms, the most significant disasters in the course of China's history have always been those affecting the densely populated valleys of the Yangtze River (Chang Jiang) and Yellow River (Huang He). Flooding on the Yellow River, often referred to as 'China's Sorrow', has probably cost more lives than any other single natural feature in the world associated with disaster. And it is probably no coincidence that El Niño exercises its strongest influence on Chinese weather in these very valleys.

El Niño's main influence is exerted on the *meiyu* – the plum rains or summer monsoonal downpour that provides the lifeblood for the hundreds of millions living along the eastern third of the country. While the inconsistent nature of the El Niño effect in China defies any simple explanation, it is fair to say that the impact is generally greatest in the north-east, north of the Yangtze River. Typically, the northern alluvial plain of the Yellow River experiences drought in an El Niño event, as it did recently in the summer of 1997, and most infamously in the Great Famine of 1877–8. The flipside of the El Niño cycle has tended unsurprisingly to result in the opposite effect, and many of the Yellow River's worst floods, but by no means all, have coincided with La Niña years.

The worst recorded flood of modern history occurred in the late spring of 1887, when between two and seven million people drowned. Another in 1954, also a La Niña year, saw 40,000 people drown, several thousand of them peasants ordered to stand with their arms interlocked and their backs against the wall of a bursting dam. Even the flood of 1938, which was rightly attributed to the deliberate destruction of dikes by the Chiang Kai-shek government in a futile bid to halt the Japanese invasion, occurred after exceptional rains, greatly contributing to the death toll of 900,000. Typically, too, the rainfall along the mid- and lower sections of the Yangtze Valley is increased during an El Niño or in the year immediately following it. Floods of magnitude are by no means confined to El Niño or La Niña years, but several of the calamitous floods in recent history do bear a relationship, such as that of 1911, which killed 200,000, and that of 1931 resulting in some 300,000 deaths. The floods of the La Niña of 1998 inflicted a much smaller death toll of 3,600, but nevertheless inundated 25 million hectares of farmland, affecting 240 million people and causing over £14 billion in damage.

China's emperors had much to fear from the consequences of natural disasters. Droughts and floods were crucial in pushing the peasants to revolt in the overthrow of the T'ang, Sui, Wang Mang[20] and, as we shall see in the next chapter, the Ming Dynasties. To the emperor, the crises of greatest anxiety were those that spread across many provinces, extended beyond more than one growing season, or followed hot on the heels of a previous calamity. Those that struck close to the capital also beckoned danger. Many characteristics of El Niño's nature spelled trouble at the top. Among these were its strong influence in the north where the emperors resided, the tendency for El Niños to be followed by the opposite phase of the cycle, and their ability to affect large areas simultaneously. The emperor's mandarins therefore had good reason to take careful note of every calamity, ever alert to the inferences that could be drawn from the standards of the emperor's conduct. These conclusions were never straightforward, and the reaction of the emperor to the subsequent plight of the people was the key to averting the outbreak of rebellion.

Emperor Dao Guang, for example, first responded to the drought that hit northern China during the 1832 El Niño by ordering all prisoners in Peking to be tried so that their guilt or innocence could be proven. In this way he hoped to right any possible injustice that may have been tampering with the harmony of the seasons. Yet, as the drought continued, he was pressed by officials to change tack and express remorse and sympathy to the people. To show he was 'mindful of their sufferings', he made this address to them:

> This year the drought is most unusual. Summer is past, and no rain has fallen. Not only do agriculture and human beings feel the dire calamity, but also beasts and insects, herbs and trees, almost cease to live. I, the minister of Heaven, am placed over mankind, and am responsible for keeping the world in order and tranquillising the people. Although it is now impossible

for me to sleep or eat with composure, although I am scorched with grief and tremble with anxiety, still, after all, no genial and copious showers have been obtained.[21]

The fact that Dao Guang first responded by seeking blame elsewhere was far from unusual. Although there are many historical examples of shamed officials offering to be put to death to remove the cause of prolonged disasters, the instances of emperors admitting personal failings are few and far between. No doubt all emperors believed in the 'Mandate of Heaven', but they were never slow to manipulate the ideology to avoid criticism. Initial responsibility was invariably sought among the government and administrators of the region where the calamity was greatest. The inclination to offset blame was balanced by the way the ruler was perceived in the countryside. Only when all initial remedies failed, or the disaster had spread across several provinces, did the eyes of the emperor's court begin to look inwards in self-examination.

In truly desperate times, when the clamour of criticism was ringing alarmingly in the ears of the royal court, offers of sympathy turned to pleas for forgiveness. 'The memorialist, his heart wrung with despairing pity', began an address to the throne by a court official during the Great Famine of 1877–8, 'cannot but ask why has a calamity so awful as this been visited upon the people. He can only ascribe it to his own failure in the due discharge of his duty, and feels that his shortcomings admits of no excuse.'[22] Other public gestures were also made. The consumption of meat in the palace was stopped, wasteful banquets were banned and public prayers were offered.

The pall of over 13 million deaths from starvation took its toll on the royal court. The Mandate of Heaven was heavily questioned following the Great Famine, and while the collapse of the Qing Dynasty did not happen immediately, its credibility was fatally undermined.

Climatic failure was not an uncommon reason for people to reject their leadership. The leaders of many empires and kingdoms owed their right to rule to their supposed supernatural status, and their ability to act as intermediaries between heaven and earth. Natural disasters were viewed as an omen, open to interpretation by the leaders and their people. Those leaders seen as corrupt, or whose actions were deemed inadequate, were vulnerable to being swept away by the masses, who believed they had lost their divine right to rule. In this way, El Niño's periodic visits challenged the legitimacy of rulers, costing many their lives.

The kings of the central African state of Jukun, a region whose inter-annual variability in rainfall is probably more affected by Atlantic sea temperatures than by the Pacific, were so identified with a successful harvest that each was addressed as 'Our Ground-nuts' or 'Our Beans'. A series of droughts or bad harvests would be 'ascribed to his negligence', and 'accordingly [he would be] secretly strangled'.[23]

Likewise, the chiefs of many small Pacific islands led precarious reigns, dotted with inter- and intra-island conflict brought on periodically by the diminution of their limited water and food supplies. The body of the last recorded *Patuiki* or king of Niue, a French Polynesian island with no natural supply of surface water, was discovered in the forest during a great famine. His death, which probably dates to the exceptionally strong 1790–3 El Niño, is still remembered and he is known today as *Ike ne kai he kuma he mate pipili*, or 'The king that was food for the rats when he was starving'.[24]

Egyptian history serves as a good reference point again. Contemporary historian Cyril Aldred described any ruler of the country as a 'rainmaker who kept his tribe, their crops and beasts in good health by exercising a magic control over the weather [and who was] able to sustain the entire nation by having command over the Nile flood'.[25]

The Nile's exceptional reliability left the masses generally inclined to believe in the Pharaohs' claims of influence over its life-sustaining floods. 'No one hungered in my years',[26] the proud boast of King Amenemhet of the 8th Dynasty, echoed through the centuries as the credit for the annual floods was eagerly claimed by those acting as demi-gods. But such claims had a downside.

In the words of Cyril Aldred, 'The king was the personi-fication of *ma'at*, a word which we translate as "rightness" or "truth" or "justice" but which also seems to have the meaning of "the natural cosmic order". The forces of evil could upset *ma' at* until restoration had been effected by some appropriate act – a magic rite, or the advent of the new king.'[27]

It is now generally accepted that one of the most turbulent periods in Egyptian history, characterised by anarchy, disorder and a rapid turnover of rulers, also coincided with a period of famine and poor Nile floods. From around 2180 BC, following the end of the long reign of Pepi II, who was responsible for one of the pyramids of Saqqarah, texts point to the mysterious disappearance of numerous successive monarchs, suggesting a desperate search for the restoration of *ma'at*. One Greek historian records seventy kings in as many years, although another historian lists a possibly more realistic eighteen kings and one queen in twenty odd years over the same period. This period, which has come to be known as the First Dark Age or First Intermediate, saw the collapse of centralised control from the Pharaohs' power base at Memphis. The winners created by this power vacuum were the nomarchs, the rulers of Egypt's nomes or mini-states. Their power was invested in an ability to divert water into their nomes and maintain sufficient food supplies. One such nomarch, Ankhtifi, who presided over the two most southerly nomes of Upper Egypt, boasted that 'the entire country had become like a starved grasshopper, with people going to the north and to the south [in search of grain], but I never permitted it to happen that anyone had to embark from this to another nome'.[28]

Rulers the world over were desperate to disassociate themselves from the impact of natural disasters. Maintaining the status quo made contingency planning for natural disasters essential in order to check their magnitude and restrict their impacts to a minimum. It was also important for rulers to instil confidence in their subjects by being seen to be securing against future disaster. For example, the need for the Incan élite to display their preparations for drought was so vital that their *qollqas*, or granaries, were among the most ostentatious and prominently positioned of all their architecture.

For rulers, rituals served to divert the perceived cause of calamity away from themselves. For example, blame for disasters could be apportioned on wider society for its in-attention to religious obligations. Rituals also provided an opportunity for the rulers to be seen to be taking action and expressing empathy with their subjects' plight, such as through prayers, acts of penance or even sacrifices. In this way, rituals underlined the wisdom and indispensability of authority, and helped to offset any groundswell of revolt. For the people, rituals helped replace feelings of impotence with a sense of control and provided a form of social cohesion through participation in collective action.

Most weather-orientated rituals were designed to meet all contingencies to guarantee their success and reinforce belief in their efficacy. For example, in times of drought, inhabitants of the Pacific's Loyalty Islands would send two sacred men to call on the spirits of a respected ancestor by collecting the bones from a grave. The entire skeleton would then be skilfully reconstructed before being hung in a nearby cave where the sacred men would douse it in water. In times of flood, water would be replaced by fire and the skeleton set alight. In either case, the sacred men would remain in the cave until the climatic fortunes had changed.

Even when adverse weather followed the performance of the appropriate rituals, fault was commonly attributed to human error, whether through sorcery, sin or incorrect performance of the ritual. The validity of the ritual itself was rarely questioned.

Societies the world over have developed rituals to prevent disastrous weather and to mitigate the calamities that have already befallen them. The communities that have shown the heaviest use and reliance on these weather-orientated rituals are those subsisting in the most unstable climatic regions. A common thread throughout these societies has been the need to communicate with the appropriate forces via a medium or gift, the ultimate offer widely believed to be the sacrifice of a human.

There are contemporary examples of a few of these ancient practices that are still conducted today. The Aymara of the Bolivian Andes, ancestors of the once powerful Tiwanaku civilisation, communicate with the mountain spirits that control the rains through the sacrifice of frogs. In times of drought, typically associated with El Niño years in the central Andean highlands, the Aymara leave frogs to dry out in the sun, hoping their dying croaks will invoke the pity of the spirits who send rain.

The belief in the power of frogs to bring rain is surprisingly widespread. Perhaps it is the frog's habit of burrowing into the earth in the dry season and re-emerging at the onset of the wet that has led many to see it as a mediator between man and the rain-spirits. When the monsoon's arrival is delayed in villages in parts of southern India, a frog is tied to a fan and placed at the doorstep of each house for the owner to sprinkle with water. When similar worries confront certain Mayan communities in central America, all villagers are required to attend a special ceremony where four boys shelter under a table from where they croak a 'frog-note' particular to the occasion.

There is much evidence in historical and archaeological records, particularly those pre-dating the influence of Christian missionaries, of humans being the preferred objects of sacrifice.

In some cases, the sacrificial victims were also those blamed for causing the rains to fail. Whenever drought struck the Pacific's Marquesas Islands, a common occurrence during La Niñas, 'fishing raids' were conducted against neighbouring islands to catch the people whose curses were said to have caused the staple breadfruit to drop before ripening. The situation on the neighbouring islands was often just as desperate, so that the raiders rarely met with much resistance. According to a missionary who had just arrived during the drought of 1797, one prisoner destined for sacrifice was already so weak that 'the natives amused themselves with giving her a slight push which was sufficient to bring her to the ground, against which her bones rattled like a skeleton'.[29]

In the vast majority of recorded cases of human sacrifice, the victims were wholly innocent. They were usually offered as a gift of appeasement – the ultimate gift that man could offer to the gods. This highest of sacrifices was generally viewed as a positive action that would preserve the life of many in return for the loss of just a few.

There are a number of examples of human sacrifices being used to avert and mitigate climatic catastrophe. In southern China, the spring boat festival which still takes place once featured a race between precariously narrow boats, that lasted until one capsized and the crew drowned. On the Yellow River in the north of the country, the most beautiful girl of the district was offered as a bride to the 'Count of the

River'. She was dressed in finery and jewels, placed on a marriage bed and launched into the river. A similar tradition took place on the Nile where only virgins were selected.

Evidence of sacrifices by the Moche society of ancient Peru suggests that some of their ceremonies were reserved for times of catastrophe. The Moche kingdom lasted from around the time of Christ to AD 750. At its peak during the fifth and sixth centuries, its influence stretched from its power base in the valley of the Moche River along some 400 kilometres of northern Peru's coastal plains. This state-level society was able to support its substantial population by exploiting two complementary food sources. Firstly, vast quantities of carbohydrate-rich grain were extracted from the dry floodplains of the equally dry river valleys and strips of coastal desert in between. These were supported by a network of irrigation canals that channelled the spring melt and rainwater that seasonally flowed from the Andean foothills. The second source of nutrition, this one rich in protein, came from the seas. Just offshore, one of the world's richest fisheries was exploited by the many fishing villages dotted along the coast. In addition, fishermen also supplied fish-meal and guano collected from nearby islands both of which were used to fertilise the fields.

However, the foundations of the Moche culture, with its cities built on the floodplains of normally docile rivers and its indispensable irrigation system, were in fact potentially vulnerable to severe flooding. Every few decades, during strong El Niño years, heavy rainfall on the northern Peruvian lowlands produces great floods that surge down the valleys. The consequences of these occasional deluges had a great impact on Moche history as will be explored in the next chapter.

It appears that the organisation of Moche society was built upon the centralised rule of priestly lords, supported by warriors and bureaucrats. The Moche era was a culturally sophisticated civilisation, known for its painted ceramics and advanced metallurgy, and above all its great adobe monuments. These structures rose sharply out of the flat, barren plains – intended as a clear reflection of the empire's wealth and the divine power of its intermediaries with the spiritual world.

The two grandest of all the great Moche monuments were the gaily-painted pyramids of Huaca de la Sol and Huaca de la Luna, the temples of the Sun and Moon. They took pride of place in the Moche Valley, the urban and ceremonial capital of the vast Moche kingdom. Today they exist as crumbling ruins, near the present-day town of Trujillo. The Huaca de la Sol was a cross-shaped structure, consisting of platforms, terraces and plazas, with a pyramid at one end thought to contain the palace of the supreme lord. In its original glory, the temple measured 345 by 160 metres at its base and stood 40 metres above the plain. More than 143 million sun-baked adobe bricks were used in its construction. The Huaca de la Sol was the largest adobe monument ever to be built in the New World, amounting to three-quarters of the size of the Great Pyramid of Giza.

The Huaca de la Luna, its slightly smaller cousin, faced it several hundred metres away. This temple is believed to have served as the home of the gods and it was here that the ceremonies and rites associated with the gods took place. On its walls multi-coloured friezes depicted a rich variety of anthropomorphic beings and supernatural scenes, many of them relating to rainfall, fertility and the perpetuation of society. One particular patio was decorated by murals of warriors and a deity known as the Decapitator, a fanged being depicted holding a knife and a severed head. Alongside these graphic scenes were more apparently peaceful ones depicting octopuses, catfish and stingrays. The connection between the two sets of images, and to El Niño events, has only recently been discovered.

During excavation work at Huaca de la Luna in 1995, archaeologist Steve Bourget unearthed the remains of over seventy bodies. These were found in a plaza believed to have been built just prior to their sacrifice, possibly at the first signs of flood. The manner in which these people were killed, and the artefacts found along with them, say much about the sacrificial ceremonies that were probably widespread throughout the Moche kingdom in times of crisis.

Judging from the degree of musculature and signs of old injuries, most of the bodies belonged to young male warriors. These warriors may have been captured in ritualised warfare, possibly in battles designed for the express purpose of providing sacrificial victims. Bourget found that many of the craniums had been smashed and numerous bones carefully rearranged. The ribs, jaws and toes of some victims had been inserted into the skeletons of others. Broken clay effigies were found between some of the bodies. Fragments of rock projectiles taken from a nearby sacred rocky outcrop were scattered around them, suggesting the effigies had been smashed deliberately. A sea lion's tooth was also found resting on one of the bodies. Nearby, in a tomb occupied by a much older man who is thought to have conducted sacrifices, a heavily worn wooden mace daubed with human blood was discovered.

Even more intriguingly, buried below the bones of the warriors were the remains of two headless children, one of them with a small whistle in each hand. From Moche iconography elsewhere, it is thought that the act of whistling was performed for children destined for sacrifice to warn their ancestors of their impending departure so that they could be carried into the afterlife, often by an anthropomorphised bat.

The motivation for the sacrifices was obvious to Bourget from the way the bones were encased in a yellowy mud, the same material used to construct the temple walls. The discovery of the sea-lion tooth provided him with a further clue to the rationale behind the sacrifices. It also became clear to him that the sacrificial ceremonies were repeated several times, each time using the same ritualistic procedures and each time involving greater numbers of human victims.

It was evident to Bourget that the scenes in the murals had actually been enacted. Slowly he began to piece together a nightmarish scenario from the evidence left to interpret. The Moche had been cursed by torrential rains and floods that caused murals to fade and turned cracks in the temple walls into chasms. Vital irrigation canals were severed, and the normally rich seas were emptied of anchovies and sardines. The Moche even noticed a change in the local sea lions that had become hungry and aggressive, attacking their fishing nets. In short, the deluge threatened to tear apart the very fabric of Moche society. Propitiatory sacrifices were called for to appease the gods, but as the rain kept falling, more were conducted. Only when the flooding ended did they finally cease.

Human sacrifice as a response to abnormal weather reached its zenith in central America. Rain cults extended across central America into southern America, and rain gods were among the most important deities. Here, rulers did not wait for calamity to strike before making their offerings.

At the beginning of each planting season, the Mayans threw a large group of captive men, women and children into cenotes or sacred wells, along with jewels and other valuables. If by noon any victims had survived several hours in the water, they were rescued and asked to reveal what rainfall Chaac, the rain god, intended to send over the coming months. If they died, as the vast majority did, Chaac was at least thought to have been appeased.

Concern over the coming year's rainfall was equally acute among the Aztecs, whose rain god was known as Tlaloc.

More images of Tlaloc have been found than those of any other Aztecan deity. Generally, Tlaloc was considered to be a benevolent god, but when aggravated he could withhold all rain or unleash devastating floods. To appease his rage, men were led to the top of the pyramids and beheaded. Women were forced to dance as goddesses of the earth before being slaughtered. And children, who were thought to represent little Tlalocs serving the chief god, had their nails extracted so that their cries could be interpreted as signs of forthcoming weather, before being drowned.

Even though the Aztec custom for human sacrifice is known to pre-date the fifteenth century, it was the consequences of a series of El Niño years around 1450, to be explored in the next chapter, that made the empire synonymous with the taking of human life. By the time European *conquistadores* arrived in the early sixteenth century, up to 250,000 victims, according to one estimate, were being offered to Tlaloc annually.[30]

Christian priests were appalled by the barbarism. Despite strong resistance from the old pagan order, human sacrifices were banned, and the idols of pagan gods were replaced by images of the Christian divinities. In the battle for hearts and minds, the new Christian order had to provide the millions of peasants with a convincing explanation for the vicissitudes of their natural world. It wanted the peasants to believe that the causes of floods and droughts lay not in Tlaloc's anger through neglect of traditional obligations, but in human sin for which they needed to repent.

The Virgin of Los Remedios carried special status in Mexico. She was the first apparition of Mary, the mother of Christ, and the conquistadores believed her statue played a vital role in Cortes's victory. When drought and flood struck after the conquest, appeals were made not to Tlaloc's beneficence through human sacrifice, but to the Virgin through

penitence and prayer. When conditions were exceptionally
severe, as they often were during strong El Niños the statue of
the Virgin was brought from her home in the west of Mexico
City to the Metropolitan Cathedral, accompanied by a procession of thousands of flagellants, religious officials and food
vendors. Her reputed efficacy in bringing rain was no doubt
partially attributable to the church authorities' timing. The
adverse conditions of an El Niño would normally last no
longer than a year, a convenient period for the authorities.
They could spend several months assessing the event before it
was deemed serious enough to transport the Virgin to the
capital, where she was left until the weather had changed. The
use of the Virgin as the ultimate saviour in the face of a
prolonged bad year gradually lost favour during the seventeenth century, as officials began to appropriate her as an
official royal image. By the time of the War of Independence,
when she was co-opted by the anti-Independence forces, belief
in her power over the weather had waned completely.

All around the world, missionaries seized on climatically
abnormal years as ideal opportunities for selling their own
message. The missionary records from Polynesia show that
exceptional weather was widely experienced during the El
Niño of 1844–6. These helped provide valuable ammunition
during a key period in the battle for Christian domination of
the South Seas. On New Caledonia, locals were taught that
'the seven plagues'[31] they were experiencing – drought, fleas,
mosquitoes, famine, epidemic, flies and a burning sun – had
been inflicted by the 'anger of the all-powerful' at the islanders' refusal to give up cannibalism and sorcery and embrace
the holy faith. On Aneityum in the New Hebrides, the rare
occurrence in the region of a devastating cyclone split the
islanders into two camps of faith. One believed that the cause
lay in 'a manifestation of Jehovah's displeasure for the continued opposition of the heathen', the other attributed it to the

'anger of the gods of their forefathers, whose right of dominion had been called in question by the Christians'.[32] Another
cyclone in 1846 brought Christianity to the islanders of
Fakaofu by destroying all their coconut trees and forcing
them to decamp to the larger neighbouring island of Uvea,
where a missionary was based.[33]

On Rarotonga, many disbelievers were also converted in
the aftermath of an unexpected cyclone by 'one of the most
devastating afflictions that ever befell the island . . . plantations, houses, chapels, were made the plaything of the storm,
and our expectations would have been realised had the
foundations of the island itself been broken up, and every
vestige of its existence been swept from the bosom of the
sea'.[34] The inhabitants were left facing months of famine. But
the disaster turned to opportunity for the local missionary
when an American ship laden with supplies docked a few days
later. The ship was in fact sailing between Sydney and Tahiti,
but was driven off her intended course by contrary winds, or,
as the reverend widely exclaimed, because the 'Lord had
brought them the right way'.

In times of adverse weather and subsequent crop failure, a
mission's wealth and superior resources became the ammunition used to break down heathen resistance. Souls were
won by offers of food that for months on end the land could
not provide. 'Proselytes come in swarms as thin as skeletons',[35] was how one excited Protestant minister described
the response during the great El Niño related famine of
1888–92 in Ethiopia. In this unforgiving climate, the opportunities for conversions created stiff competition between
missionaries, and goodwill between the rival faiths became
as scarce as the rains. 'The Abyssinian is in poverty,' wrote a
jealous Father Crombette during the same famine. 'Beside
him the Protestant displays his gold, but he does not give it
to him except on condition of selling him his soul and

abandoning it to him for eternity.'[36] Commonly, to the missionary's regret, abandonment of the soul lasted only until the next big rains.

When the Baptist missionary Timothy Richard first realised the opportunity presented by China's 'Great Famine' of the 1877–8 El Niño to exercise influence over millions of people, he confessed: 'A powerful thrill affected me so that I could hardly walk.'[37] However, the missionary's ability to offer food in China's time of crisis was met with less than overwhelming approval. Anti-imperialist authorities in China were highly suspicious of Richard's agenda and that of other missionaries. As a result, some foreign relief programmes were banned and rumours were spread that the foreigners 'were taking in orphans for the purposes of scooping out their eyes and cutting out their hearts for various alchemic purposes'.[38]

During the simultaneous famine in India, high-caste Hindus also objected strongly to the missionary-based relief efforts that threatened to create 'rice-Christians'. Meanwhile in Vietnam, during major flooding and a related cholera outbreak, apostles of the Dao Lanh sect capitalised on the peasants' misery by inciting revolt against the French occupation and distributing amulets that supposedly offered protection against French arrows and bullets. 'The harvest is decidedly bad,' wrote the French inspector at My Tho. 'Some ill-disposed people have been trying to take advantage of the fear of famine to incite people to rebellion.'[39] After the planned uprising was foiled by a spy, and the complicity of some Buddhist monks exposed, the French prohibited any further construction of religious buildings without authorisation. The law remained in place until the Vietnam War.

4

Reign of Terror
Conflict, Famine and Migration

No rain comes from Heaven,
The earth is parched and dry,
And all because the churches,
Have bottled up the sky.

Boxer Rebellion manifesto, inciting anti-Christian sentiment
during a drought in Northern China, 1899
Tao Teh Ching, Lao Tzu[1]

When archaeologists first began to excavate the Moche
temples of the Sun and Moon in the early 1970s, they believed
these isolated monuments set in a desolate landscape must
have serviced a distant population. When digging started,
only a few urban buildings were visible, but excavations
between the two temples revealed a vast city, entombed at
the height of its powers in several metres of sediment.

Some time around AD 550, a cataclysmic El Niño struck –
an event of such severity that only one of its magnitude can be
expected on average every 500 years. Exceptional rains
poured on to the Peruvian lowlands, and torrential waters
surged down the Moche River, stripping several metres of soil
from the valley floor. The water also scraped away the base of

the temples or *huacas* and levelled many smaller buildings to the ground. Adobe houses crumbled under the deluge. In all probability sacrifices of warriors were made at this time to call on the spirits to stop the rain. The core of the city remained and the Moche set about repairing their capital in the ensuing years, adding extra platforms to the *huacas* and pasting on new façades to obscure the flood damage. On one new façade a colourful mural depicts a figure holding a staff in each hand, a figure archaeologists associate with the influence of the Huari empire from the Andean highlands. In addition, the discovery of a new strain of maize introduced from the Andes provides further evidence of highland influence.

But just as memories of the killer flood were beginning to fade, a new and more insidious destroyer was encroaching from the coast. Gradually, over the next few decades, winds blew great ridges of sand inland. Irrigation canals were clogged and sand dunes enveloped parts of the city. For many years, the Moche must have fought the untimely invasion of their thriving metropolis. But gripped by a prolonged drought between AD 562 and 594 in the Andes, the source of the rivers essential for irrigation, they were powerless to halt the advancing sand. Finally beaten, they removed roof beams along with other valuables, and abandoned the perfect shell of their city to an early grave.

At first glance, this second, fatal blow to the Moche city appeared to archaeologists to be unrelated to the flood. But evidence soon pointed to an El Niño connection. The first clue arrived when the 1983 space shuttle *Challenger* transmitted a satellite image of the Santa River region, fifty miles south of the Moche Valley. Upon analysis, it was noticed that a new ridge of sand had formed near the river mouth, a dune that had not been there fifteen years before. How had it got there?

On 31 May 1970, an earthquake registering 7.9 on the Richter scale struck northern Peru. With a death toll of over

75,000, it remains the most devastating natural catastrophe in the New World during the last 500 years. Accompanying this disaster were thousands of landslides, including one in the Callejon de Huaylas that killed 18,000, the deadliest avalanche in history. For two years, millions of cubic metres of liberated debris remained in a sort of suspended animation, until heavy rains could wash them out to sea. The El Niño of 1972–3 brought torrential rains to Peru, causing flash floods that severed all northern highways for months on end and washed away whole adobe villages. Much of the debris was flushed into the sea. Within two years it was already apparent that wave action had formed identifiable new dunes in places along the seashore, a process boosted by the violent seastorms that characterised the El Niño year. In time, offshore winds could dump these dunes on to the heavily irrigated fields of the adjacent Peruvian coast.

The Nazca fault-line runs virtually the length of the west coast of South America. Combined with El Niño's periodic upheavals, this makes the Andean coastline one of the most dynamic on earth, with tales of incredible destruction littering its history. An offshore earthquake in 1746, for example, levelled Lima and generated an eighteen-metre high *tsunami*, or tidal wave, which obliterated the city's port of Callao. In 1835, an earthquake levelled the Chilean city of Concepción sweeping its port out to sea. By coincidence, Charles Darwin felt the powerful tremor as he was lying down in some woods along the Chilean coast, during a break in the *Beagle*'s voyage. When he finally reached Concepción a fortnight later, he found: '. . . The ruins were so confused and mingled, and the scene has so little the air of a habitable place, that it is difficult to understand . . . their former appearance or condition.'[2] Each millennium, hundreds of earthquakes rattle the region, and dozens of strong El Niños bring violent rains. With statistical inevitability, the occasional strong El Niño event will follow hot on the heels of an earthquake. Whenever this double action occurs, the potential exists for a major re-sculpting of the landscape. It is not hard to imagine that the two-step disaster that struck the Moche capital in the sixth century is another example of these two Peruvian 'horsemen of the apocalypse'.

After abandoning the site of the two *huacas*, residents of the valley moved many miles upriver to the neck of the flood-plains where a new, smaller city of Galindo was established. At roughly the same time, the Moche inhabitants of the Lambayeque Valley also shifted their city upstream, to a new site at Pampa Grande, where another monumental *huaca* was constructed. Dominion over the river valleys south of the Moche valley was lost, and centralised control of the state moved from the Moche Valley north to the new capital of Pampa Grande. There were many other changes in the new phase of the Moche empire. Archaeologists have noted a higher concentration of people within the city's boundaries, new religious iconography and tighter centralised control of the bean and maize stores. More significantly, the credibility of the people's cosmological order may have been fatally undermined. When the next great flood struck around AD 750, the inhabitants could take no more. Pampa Grande's new *huaca* was torched, possibly by its own people, and the cities of Galindo and Pampa Grande were abandoned, her-alding the final disintegration of the Moche empire.

The role of climate in history is an issue few historians agree on. Around the turn of the twentieth century, a popular school of thought espoused the view that climate had exerted an immense influence on the course of history. Ellsworth Huntington, for example, one of the leading proponents of this theory, believed climate to be 'one of the controlling causes of the rise and fall of the great nations of the world'.[3]

Over the years, such views lost credibility, partly because they were used by some to justify Western superiority. Since then, many historians have even been reluctant to afford climate any significant role in the shaping of world events. And compared to other historical influences, year-to-year climatic variation has been seen as too peripheral and short-term to be of consequence. The contemporary economic historian David Arnold has noted a scepticism 'that anything so momentous as human history can be shaped by anything so commonplace as weather'.[4]

The last years of the twentieth century have seen a sea-change in the interpretation of historical events. In the past, while geographers have looked at climate most historians have attributed the catalyst for change to man-made forces. Recently, there has been a shift in emphasis towards a more integrated, ecologically based approach that has highlighted the role of natural disasters, particularly drought, in historical change. There is now a general consensus that prolonged drought on a decade or century-long scale has been a significant factor in the decline of many civilisations, such as the Harappan of north-west India around the time of Christ, the Mayan between AD 750 and 900, and the Tiwanaku of the Andes around AD 1100. Recent analysis of archaeological and documentary evidence is now validating the role of climatic fluctuations of a much shorter timescale. While El Niño has not acted as the sole agent for change, its fluctuations have arguably exerted a subtle but profound influence on many significant historical events.

El Niños and La Niñas have struck like an irregular heart-beat throughout history, each event resulting in anomalous weather in some part of the world. The pulse of most events is too weak to be felt far from the tropics, and is neither strong enough nor extended enough to overcome the cultural adaptations that have evolved to cope with year-to-year climatic variability.

Yet throughout history, there have been infrequent beats of such severity or of such ill-timing that they have surpassed society's capacity to deal with them. In such times, the impacts have claimed untold millions of lives, triggered wars, changed ideologies, shifted populations, and ended whole civilisations.

Determining El Niño's exact contribution to such events is impossible for two reasons. Firstly, there is a need to assess what stress people may already be facing from other sources such as political and economic stagnation, or unrelated natural disasters. Even the impact of the AD 550 'mega-niño', which destroyed the two *huacas* and marked the start of the Moche kingdom's decline, cannot be easily separated from the stress of the forty-year drought. A year-long climatic failure imposed on an already weakened state can act like a bout of pneumonia on a patient reeling with flu.

Secondly, it is generally impossible to prove El Niño's role in any particular short-term weather-related anomaly. Although the El Niño system is the world's greatest source of climatic fluctuation on a year-to-year scale, there are still other important oscillations within the global atmospheric circulation, particularly outside the tropics, as well as natural random variations.

One region where archaeological evidence has afforded a reasonable interpretation of El Niño's impact is the Peruvian coast. Here years of major flood damage are virtually synonymous with strong El Niño events. Research over the past thirty years has drawn a picture that suggests the devastation and cultural upheaval experienced by the Moche have occurred many times in the past.

One episode of flooding, dated to the El Niño of 1607, came soon after two huge eruptions, one of them, at Huaynaputina in 1600, the biggest in Peruvian history. The combination of these disasters led to the partial destruction of agriculture in the area. A great flood, carbon dated to around AD 1360, caused the population in one valley to drop by over 80 per cent. Another, dated roughly around 800 BC, has been linked to the sudden rise of Chavin de Huantar, the home of the first civilisation of the central Andes and the pilgrimage centre of a highly influential religious cult.

According to a myth told to the Spanish *conquistadores*, it was around the time of the Moche's decline that a ruler called Naymlap arrived by sea with his wife. Naymlap established his capital at the city of Chot, where an idol of green stone was displayed for his subjects to revere. He also founded a dynastic line of kings, which ended with Fempellec, the twelfth and last. Fempellec is reputed to have been tempted by a sorceress to move the green idol, triggering thirty days of devastating rains and floods. After an ensuing famine and pestilence, Fempellec's subjects rose up against him and cast him into the sea, bound hand and foot.

Although this story has yet to be authenticated by archaeologists, its timing can be reasonably dated according to the known chronology of the Naymlap kings – and the date suggests Fempellec may have succumbed to a mortal blow from the strongest El Niño of them all.

By the beginning of the twelfth century, two-thirds of Peru's desert coast fell under control of the Chimu state. Known for their distinctive pottery, textiles and metalwork, the Chimus' greatest achievements were arguably their feats of engineering. A vast, centralised network of irrigation canals crisscrossed the coastal plains, sometimes stretching over a hundred kilometres between river valleys, to reclaim the inhospitable desert for agriculture. The surveying expertise needed for their construction was phenomenal. Even with modern technology, the design and completion of such a complex network of canals to supply an optimal flow of water would be extremely challenging.

Around AD 1100, this extraordinary engineering feat was destroyed in a single flood. With floodwaters over eighteen metres high in the Moche Valley, aqueducts were levelled, roads disappeared, and several metres of the valley floor were stripped away. Even the destructive floodwaters of 1925, the greatest since the Spanish Conquest, fell some 10–15 metres short of the Chimu deluge.

Around the same time, other cities also appear to have experienced catastrophic flooding. To the immediate north of the Moche Valley, the Chimu capital of Chan Chan, as well as Pacatnamu, were affected. Further north again in the Lambayeque region, the flood damage to two cities of the neighbouring Sicán state – Chotuna and the religious centre of Batan Grande – left both significantly weakened. According to the Naymlap legend, Chotuna, once probably the city of Chot, was abandoned.[5] Batan Grande suffered a similar fate after its inhabitants stacked great wooden piles against its temples and monuments and let the city burn. Sicán religion brought in new beliefs and iconography, and the capital was transferred to a new site at Túcume, signalling an even greater phase in the construction of monument building. Túcume's twenty-six adobe pyramids became the greatest complex of such structures in the Americas.

The Chimu on the other hand responded to the disaster by greatly accelerating their military campaigns against their neighbours. With the flood permanently destroying their irrigation network, and much land being lost for ever to production, only the conquering of new agricultural land to the north and south could ensure the survival of the Chimu empire. Eventually, remote states such as the Sicán were subdued and annexed, and the Chimu empire grew to such an extent that it would eventually challenge the other great Peruvian power of the era, the Incas.

The history of Peruvian civilisation symbolises the theory of punctuated equilibrium, first proposed by the biologists Stephen Jay Gould and Niles Eldredge to describe the progress of evolution as the pattern of long-term stasis within species, followed by the rapid evolution of new species. Peru's story has largely been one of long periods of cultural continuity broken by rapid bursts of change, abandonment and reconstruction. In some places these changes have involved attempts at rebuilding the cities and irrigation systems. But in many instances, the destruction of the economic base either forced abandonment and resettlement elsewhere within their own valley, or the formation of new political and cultural links further afield. Even today, over three million hectares of once-cultivated land lie fallow along the Andean coast between Colombia and Chile, a testament to these profound cycles of abandonment.

El Niño events occur frequently, on average every four years or so, and even strong events recur with a frequency of about every fifteen years. But very few result in crisis, proof of the fundamental resilience of societies to cope with abnormal fluctuations. So why should one particular El Niño have more impact than any other?

The primary answer is rooted in the social, political and economic conditions that prevail at the time. El Niños have acted like historical markers, exposing fundamental weaknesses and injustices in societies and their systems. In the case of the Chimu empire, food production was overly dependent on vast irrigation systems, which no matter how well maintained and fortified were defenceless against the onslaught of exceptional El Niño floods.

But the nature of the El Niño itself is also an essential factor. The stronger the event, the more innate power to expose the weaknesses in a state and effect lasting change. The strongest events have been dubbed 'mega-niños', a term first introduced by archaeologists to capture the truly exceptional nature of

the events that have left an indelible mark on Peru's archaeological records. How the powerful El Niños of 1997–8 and
1982–3 compare in strength with these 'mega-niños' is difficult to judge. But based on the degree of flooding in northern
Peru, they are dwarfed by such predecessors.

Archaeologists have not agreed on a clear definition for a
'mega-niño'. One study based on Peruvian flood deposits
estimates an average mega-niño frequency of one every 400
years. Others using a different definition imply a higher frequency. Others again believe that a mega-niño describes not just
a single event but a series of closely timed strong events,
accumulating catastrophic impacts. But there is a danger in
extrapolating the damage in one region to assess the strength of
the El Niño in another. Local and regional meteorological
conditions can greatly alter the severity of the climatic abnorm-
ality. An interesting advancement in the understanding of mega-
niños has recently come from archaeologists across the Americas who are detecting profound cultural changes that corre-
spond to the Peruvian dates. If substantiated, this research
would suggest that other archaeologists may uncover evidence
in the future of cultural changes far from the Americas that
coincide with the mega-niños of ancient Peru.

The Amazonian basin may seem an unlikely place to look
for evidence of mega-niños. It is hard to imagine its seemingly
non-flammable rainforests suffering prolonged drought, but
analysis of Amazonian soils has revealed that drought and fire
have been regular, if not frequent, visitors. During a severe El
Niño drought in 1926, a great fire blazed for over a month in
the Rio Negro catchment. Major incidents have also been
dated to around AD 500, 1000, 1200 and 1500. These dates
are synchronous with discontinuities in the archaeological
records taken from pottery evidence. And taken together with
the extreme heterogeneity of languages and genes among the
region's people, this evidence suggests repeated episodes of

dispersal. It appears that on several occasions, semi-sedentary populations were unable to cope with the exceptional conditions and were forced to disband in search of food, later forging new alliances with other communities.

In the state of Veracruz on Mexico's central Gulf coast, similar dates mark the end of two important cities after long periods of stability and prosperity. El Pital was a significant trading and cultural outpost of Early Classic Civilisation in Mesoamerica. A contemporary of Teotihuacan in central Mexico, this vast metropolis, supported by massive raised field irrigation systems, was eclipsed suddenly some time after AD 500. Emerging soon afterwards from El Pital's shadow was the previously small settlement of El Tajin, about sixty kilometres to the north. El Tajin rose to become a powerful militaristic centre, dominating the coastal plain and probably much of the adjoining inland area too. This city, with its own distinctive architectural style, was very influential in Mesoamerica's Classic period. The ball-game cult may even have originated here. Then around AD 1100, El Tajin was suddenly abandoned and its people migrated southwards. By the time the Hispanic forces arrived, the site was no bigger than a small village.

Archaeologist Jeffrey Wilkerson has found the dates of AD 500 and 1100 coincide with evidence of drought and great flooding in Veracruz. And similar catastrophes have been dated to around AD 1540, as well as to 400, 1000, 1600, 2400, 4800 and 5600 BC. Based on his personal experiences of El Niño impacts in Veracruz during the 1990s, Wilkerson believes that these historical mega-niños represent the collective product of more than one El Niño event. Only prolonged stress, extending beyond the timescale of a single El Niño, would be sufficient to undermine a society's ability to continue. The mega-niños, in his opinion, are more likely to be the result of a close-knit sequence of El Niños and La Niñas that has reaped an accumulative and catastrophic impact. As an example of this, Wilkerson cites the delayed struggle experienced by his neighbouring Indian communities after the 1995 hurricane-driven flood of the Nautla River, one of the greatest in the region's history. It was only when this disaster was compounded shortly afterwards by the El Niño of 1997–8 and the La Niña of 1998–9 that the communities began to suffer real hardship. The combination of El Niño activity led to five consecutive growing seasons failing through a mixture of droughts, fires, floods and finally insect infestation. Wilkerson also supports his theory with the observation that the 1995 flood resulted in a 10–15 centimetre thick layer of sediment in the valley. In comparison, the layers laid down during the mega-niños of AD 500 and 1100 were over a metre thick.

One of the main tools to help us make sense of El Niño's historical impact is a chronology of events, otherwise known as a time-series. Various versions exist, each using different criteria to interpret the pattern over time and each resulting in minor differences in chronology between each other. Even where good meteorological data is available, disagreement can still exist over whether certain years actually constitute an El Niño or La Niña event. Further disagreement and confusion also persist over the exact timing of their development and decay. Two factors in particular add to the uncertainty.

An event's 'strength', normally defined today by the degree of sea-surface temperature change in the eastern Pacific, does not necessarily correlate to the severity of impact felt in any particular part of the world. However, 'strong' events do generally have a much greater influence on the world's weather systems and are consequently much likelier to have generated a series of world-wide climatic-related disasters in the historical records. It also follows that the 'weaker' the event, the harder it is to find the corroborating evidence to confirm its existence and timing.

The other source of uncertainty relates to how far back in time the event occurred. Even with some of the weaker events of the twentieth century, the various sources of evidence have been somewhat contradictory and inconclusive. However, a reasonably well-corroborated time-series has been developed back to the mid-nineteenth century, when instrumental records began. Before that time, the further back the event, the fewer sources of evidence exist to support it, and the less confidence there is in identifying a particular El Niño year.

The chronology of events used in this book has been based largely upon the pioneering work of American climatologist Bill Quinn. Focusing his research primarily on the west coast region of northern South America, Quinn and his various co-workers scoured the records, diaries and logs of sailors, explorers, naturalists, and even pirates, as far back as the early sixteenth century. The identification of an El Niño event was judged according to descriptions of heavy rains and flooding, supported by other information such as sailing times, sea temperatures and unusual natural phenomena.

Since establishing his original chronology in 1978, Quinn along with other researchers has sought corroborative evidence from regions other than the Peruvian coast. Correlations have been extrapolated from varied sources of data both inside and outside of the tropics. These sources have included a variety of rainfall and temperature records, as well as oddities such as fishery catches, shipwreck lists and even a Japanese shrine's records on the freezing of Lake Suwa. Quinn's last chronology, published in 1992, incorporated more data of climatic anomalies from a much wider geographic range than South America alone. This version is therefore a chronology of El Niño's global activity. By including analysis of Nile flood records, Quinn also attempted to extend the series back past the sixteenth century to AD 622. However, this extension was based on a very limited number of data sources, and without further corroborative evidence must be interpreted with caution.

Future documentary research is sure to modify and refine this chronology. Non-documentary sources of evidence, such as coral bands, tree rings and ice cores, are being analysed to see if they represent accurate reflections of climatic fluctuation and a correlation to El Niño activity. Already an analysis of two ice cores from the 18,700-foot high Quelccaya glacier in the Bolivian Andes has shown that year-to-year fluctuations in rainfall are evident from the layering of the ice cores. This promising line of research has uncovered the patterns of both long- and short-term climatic cycles over the last 1,500 years. More specifically, the ice evidence has provided a signature for the El Niño-related droughts that have struck the altiplano region around Lake Titicaca.

Equipped with a time-series and mounting documentary evidence, it is now possible to shed some light on El Niño's influence on specific events over the last millennium. This chapter provides a snapshot of El Niño's legacy on history, and some of the momentous changes it has prompted among peoples, states and nations.

1200–1[6]

One of the earliest major El Niños for which there is a detailed written account occurred in AD 1200–1. Records exist of great drought and famine in southern India and central China, but the diary of one of the most esteemed medieval physicians and scholars suggests that the people of Egypt endured the worst suffering.

Abd al-Latīf al-Baghdādī first heard of a famine in Egypt when tales reached him in Iraq of thousands emigrating from the Nile valley into Arabia, Yemen and Syria. His first-hand record, compiled after travelling to Egypt to see the conditions for himself, provides a uniquely detailed description of the famine.

Abd al-Latīf witnessed a mountain of over 20,000 corpses, 'more than the eye could perceive'. The road to Damascus was like 'a vast field sown with human corpses: or more like a country where there had passed the sickle of the reaper. It had become like a banqueting room for birds and wild beasts.'[7] There are no accurate figures for the numbers that died, but there are a few details that give some indication of the extent of the tragedy. Of the 900 weavers of rush-mats in Misr, for example, only fifteen survived. And in Alexandria, a single inheritance reputedly passed through forty heirs in two years. Over 110,000 dead were formerly presented to Cairo's cemeteries alone, but judging from reports, many more corpses never made it to burial.

Cannibalism became so rampant that those walking the streets were caught by hooks let down from windows above, tradesmen were lured to fictitious jobs only to be slaughtered while they worked, physicians consumed their patients, and hosts devoured their own friends after inviting them to dine.

There was no shortage of horrific stories to record in Abd al-Latīf's chronicle. 'One night after the sunset prayer, a young slave played with a newly weaned child which belonged to a wealthy person. While the child was at her side, a beggar, seizing a moment when the slave had her eyes turned from him, slit the child's stomach and began to devour its flesh raw.'

And Abd al-Latīf had equally grim personal observations of his own to report. 'I myself saw a small roasted child in a basket. They carried it to the Emir and led in at the same time the mother and father of the child. The Emir sentenced both of them to be burnt alive.' But Abd al-Latīf noted even this punishment did not deter the masses, because the body of one cannibal, put on public display after being burnt for his crime, was found 'devoured the next day: the people ate it very willingly, for the flesh having been well roasted did not need further cooking.'

For Abd al-Latīf the famine provided a once-in-a-lifetime opportunity to undertake certain medical experiments. With overwhelming numbers of bodies piling up in the cities, the Islamic law forbidding the dissection of bodies had been relaxed. As a result, Abd al-Latīf was able to perform the first known autopsies, making noteworthy contributions to the science of anatomy with his descriptions of the lower jaw and sacrum.

To the Christian Zealots of the Fourth Crusade, an Egypt weakened by famine presented a unique opportunity to capture a cornerstone in the Islamic empire, a conquest likely to force the cession of Jerusalem. The Europeans were already aware of Egypt's military weakness in the wake of a severe failure of the Nile. In AD 969, the Fatimids of the Tunisian region had been met with little resistance when they invaded Egypt two years after a failed Nile flood had devastated the country, killing 600,000 people around the capital of Fustat alone, the equivalent of a quarter of the population. Now in 1202, the opportunistic Crusaders set out to invade Egypt. They reached Venice but, unable to raise sufficient funds for their expedition, were persuaded by the Venetians to attack Constantinople and the Hungarian city of Zara instead. The Fourth Crusade never reached Al-Latīf's famine-stricken Egypt.

1450[8]

In the mid-fifteenth century, Emperor Moctezuma I acceded to the throne of the Aztec empire, and initiated what is generally recognised as its golden age. However, his reign began at a time of terrible harvests and a famine which changed the course of Aztec history.

In 1450, early autumn frosts in the valley of Mexico kicked

off four successive years of crop failure. Following hard on the heels of devastating floods in the Aztec city of Tenochtitlan and a locust plague throughout the valley of Mexico, the sudden cold snap was particularly cruel. 'The snow that fell over the land was so high that it reached in most parts a stadium and a half, which caused the destruction and collapse of many houses as well as the destruction of all trees and plants, and the country was so cold that many people died and especially older people,'[9] recorded an early seventeenth-century historian. The year 1451 also brought crop-destroying frosts and snows. Both years corresponded to low Nile floods, prompting a reasonable assumption that they were also El Niño years. In fact, the Nile level in 1450 was the lowest for over 300 years, and by the following year it was possible to wade across it in some places. In 1452 and 1453 the Aztec empire was gripped by drought, suggesting a change to La Niña conditions. Fountains, rivers and springs dried up, and according to a petition to Tlaloc, even the birds collapsed through thirst: 'They topple over and lie prostrate on their backs, weakly opening and closing their beaks.'[10]

At the time, the empire was based on the Triple Alliance, the union of the three leading states of the Valley of Mexico – Tenochtitlan, Texcoco and Tlacopan. Whenever famine struck, the Aztec people turned to their emperor in the expectation of receiving surplus stores of grain. But by 1452, all the royal granaries had been exhausted. According to one historical source, '. . . there was not enough to go around and for this reason some people simply made holes in the ground and crawled into them to await death and when it came they were devoured by buzzards because there was no one to bury them'.[11] After ordering his subjects to attend one last banquet, Emperor Moctezuma was forced into the humiliating proclamation: 'O my children, you know that I have done everything in my power and that the foodstuffs have been used up. It is the will now of the Lord of the heavens that each of you go his own way to seek his own salvation.'[12] Tens of thousands fled the kingdom, many dying in the forests as they scavenged for roots and tubers, and many selling themselves in return for maize to the Totonacs and Huaxtecs on the Gulf coast, an area that had not been affected by the adverse weather.

Humiliated and frightened by their sudden loss of power and people, the sovereigns of the Triple Alliance set about boosting food supply and consolidating their power. Firstly, military campaigns conquered vast swathes of territory, including the east coast states that had prospered during the famine, establishing Aztec control over most of the Meso-American population. Secondly, a massive construction programme of aqueducts and irrigation canals was undertaken to control salinity levels and promote irrigation in the Valley of Mexico. The additional produce that was cultivated helped unify the various states and cities of the Valley of Mexico more than ever before.

The third response served the dual purpose of reducing the number of food consumers while addressing the perceived cause of the problem. The adverse weather was believed to result from the retreat of the rain god Tlaloc into his cave owing to an insulting shortage of human sacrifices. Before this time, while human sacrifice had been practised, it had not been performed in any systematic way. Now, in order to avert the wrath of Tlaloc, mass offerings of thousands were deemed necessary. Fear ruled the countryside, as men, women and children were taken as tribute payment to the Aztec kings. How these sacrificial victims were chosen is not known, but it is likely that the numbers of 'innocent' citizens offered by each state were kept within limits to avoid a groundswell of revolt.

The solution was, as one historian described, 'one of the most extraordinary compacts in the world's history'.[13] By mutual agreement of all the lords and kings concerned, the

Right Passengers on board a steamer look out onto the icefield close to where the *Titanic* sank. Exceptional numbers of icebergs had drifted much further south than normal in 1912.

Below On their expedition to the South Pole in early 1912, Captain Scott and his team experienced exceptionally severe weather and low temperatures, contributing greatly to the exhaustion and frostbite that led to their deaths.

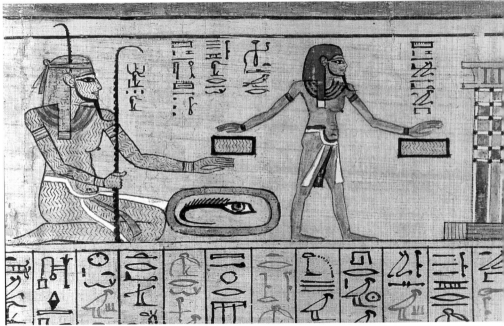

Above The River Nile's annual flood, on which the prosperity of the Egyptian civilisation depended, was believed to be controlled by the god Hapi. She is depicted here with pendulous breasts and a corpulent body from a detail of the papyrus of Ani.

Left The practice of human sacrifice by the Aztec hierarchy to feed their rain gods increased exponentially following an El Niño-related famine in 1450–4.

Above The Moche 'Temple of the Sun' of coastal Peru, was abandoned after a series of disastrous floods, evident in the erosion damage still visible at its base.

Middle right In the adjacent 'Temple of the Moon', the bones of over seventy bodies were discovered. The bodies had been sacrificed in order to placate the rain gods during a period of torrential rains sometime between AD 550 and 700.

Below right An octopus figure found on a nearby mural, in front of which the victims were probably paraded before being sacrificed.

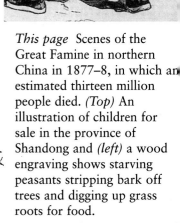

This page Scenes of the Great Famine in northern China in 1877–8, in which an estimated thirteen million people died. *(Top)* An illustration of children for sale in the province of Shandong and *(left)* a wood engraving shows starving peasants stripping bark off trees and digging up grass roots for food.

Opposite page *(Top)* Indian famine victims close to death following the monsoon failure of 1877–8, one of many climatic disasters resulting from the deadliest El Niño on record. Approximately six million people died in India, often overwhelming efforts to dispose of the corpses *(below)*.

An artist's impression of a cyclone that hit the island of Hikuéru in the Pacific's Tuamotu Archipelago killing five hundred people who had gathered on the island for the annual pearl harvest in January 1903. The region rarely experiences cyclones outside of El Niño events.

In the early hours of March 15, 1720, the commercial centre of Zaña on the northern Peruvian coast was washed away by heavy flooding, killing over three thousand residents. The height of the floodwaters is indicated on the eroded lower walls of the convent church of San Agustin, the only building that survived the flood.

Many of the major conflicts between colonists and African chiefdoms during the nineteenth century followed El Niño-droughts. *(Above)* An engraving depicts an ambush in the War of the Axe of 1846. *(Right)* A detail from Christopher Fripp's oil painting of the 'Battle of Isandlwhana', during the Zulu War of 1879.

Vietnamese peasants driving out the French authorities, following crop failure in 1930.

Northern Europe experienced the coldest temperatures for over two hundred years during the winter of 1941–2, debilitating Hitler's troops in Russia. Machinery and armaments were rendered useless in the freezing conditions, and almost as many troops succumbed to the cold as they did to the enemy action.

Triple Alliance and neighbouring cities and states began an ongoing series of staged conflicts known collectively as *Xochiyaoyotl*, or the War of Flowers. Rather than being conquered by the Triple Alliance, the non-Aztec states were allowed a form of independence on condition they participated in the War of Flowers. Its purpose was to maintain a continual supply of sacrificial warriors, and its battles served not to kill, but to maim and take prisoners. Soldiers risked paying the ultimate price because battles provided opportunities for wealth and status. With priests acting as referees on specially prepared battle arenas, armies of elaborately attired warriors fought till they fell wounded 'like a rain of blossoms'.[14]

To everyone concerned, these staged battles were seen less as wars than as marketplaces where the goods on offer '. . . constitute warm food which has only just left the oven tender for our God', as an Aztec nobleman put it. '. . . This war shall be so organised that our aim will not be to destroy these cities, but that they shall always remain standing so that whenever we might wish to do so and whenever our God might desire to eat and to enjoy himself, we can come there just like anyone who goes to the market in order to buy food.'[15] At great ceremonies, attended by the leaders of the opposing armies, the prisoners would be fed to the gods. The mass sacrifices of soldiers and others are estimated to have added an astonishing 15 per cent to the natural mortality rate.

In the end, the Aztecs' experiences in the War of Flowers played an important part in their downfall. When Hernán Cortes arrived decades later, he found much-needed refuge and extra manpower simply by manipulating the humiliation of states like Tlaxcala that had been forced to pay tribute and men to the Triple Alliance. Worst still for the Aztecs, the mock battles had instilled an approach to warfare that was futile against a force from another continent that declined to play by the same rules.

1640–1[16]

By 1640, the Ming Dynasty was on the point of paralysis. A state of near anarchy reigned in the countryside, as ruinous taxation and a series of poor harvests reduced most peasants to the brink of starvation. A corrupt and indisciplined army exacerbated ill-feeling towards the government. Almost bankrupted by several serious uprisings in the preceding decade, China had fallen under military threat from Japan, Russia and, most ominously, the Manchus. At this vulnerable point in time, the Ming Dynasty was struck by the El Niño of 1640–1 which brought with it drought, disease and locust plagues across central and northern provinces.

The circumstances of 1640 were also significant for Li Zicheng. Li was the head of a powerful bandit gang responsible for a series of successful raids in the provinces of Honan and Shensi. Sacked from his job as a postal worker, Li had chosen to join his gangster uncle in the Shensi mountains following a famine in 1631. A renowned guerrilla leader, Li organised a series of successful raids into neighbouring provinces, becoming ever bolder in his targets and venturing ever further from his base in Shensi. Deteriorating conditions in the countryside transformed Li's ambitions from that of a bandit to an emperor.

Eyewitness accounts of the time describe scenes of appalling degradation and crime. 'There was a great drought. The course of the Yellow River dried up. Vagrants filled the roads and people ate each other,'[17] recorded one local chronicle. A farmer's proverb of the time declared: 'It makes more sense to eat one's father, elder brother or husband so as to preserve one's own life, rather than have the whole family die.'[18] The more personal story of a shopkeeper gives some idea of the little value attached to human life.

> This man and his wife are dying of starvation, every day they come and beg me for a little sesame oil to keep them alive. Now the man is trying to sell me his wife. But in my house there are already more than ten women that I've bought, so what does one more matter to me. If she's cheap, I'll make a deal: if not, that's that. It's really ridiculous that he should go on bothering me like this.[19]

Markets for human flesh were opened, and even good friends no longer felt able to trust each other when walking together in the fields.

The years 1640–1 also marked the beginning of a terrible epidemic, of equal proportion to the devastating outbreak of 1588–9, for which it still vies as the most lethal epidemic of any disease in China's history.[20] The disease – the nature of which cannot be definitively established but was possibly the plague – even struck Peking which rang day and night to the constant beating of brass implements to ward off evil spirits. As will be discussed later in the book, the correlation of the two outbreaks with two very strong El Niño events suggests that even the epidemiology of the mystery disease was somehow associated with the unusual climatic conditions.

Hot on the heels of the famine, Li marched into Honan province with his bandit army, its numbers swelling by the starving masses. Resistance from the dispirited local army was minimal, allowing important cities to be captured without a struggle. Fame followed Li's trail as stories spread of how he had dismembered and eaten the Prince of Loyang to show his disgust at those who lived in luxury while the peasants starved. His reputation was no doubt enhanced by the practice of diverting a percentage of his spoils to help famine victims. In 1644, Peking fell to his troops and Li proclaimed himself emperor. But he was not there for long. At the behest of certain Ming generals, the Manchus entered the city from the north, driving Li back into obscurity. However, the demise of the Ming era was complete and the Manchus remained to proclaim a new dynasty.

1769–70[21]

Since 1765, the British East India Company had held the rights to collect revenue in Bengal, without assuming any responsibility for administration. Four years later virtually no rains fell in the provinces of Bengal and Bihar during the summer and autumn monsoons. Despite the growing inability of the peasants to pay taxes, the Company insisted revenue be 'violently kept up to its former standards'.[22] Pouring on the misery, speculation by Company representatives in the buying and selling of diminishing stores helped grain prices to sky-rocket.

As the drought wore on and the daily intake of food was reduced to one meal a day, then half a meal a day, and then less again, all normal cultural constraints broke down. 'Tender and delicate women, whose vests had never been lifted before the public gaze, came forth from their inner chambers in which eastern jealousy kept watch over their beauty, threw themselves on the earth before the passer-by, and with loud wailings, implored a handful of rice for their children.'[23] When that failed, the children were offered for sale, until no more buyers could be found. The Company, still unwilling to accept any moral obligation towards the people's welfare, 'magnanimously sanctions a grant of ten shillings' worth of rice per diem'.[24] This pitiful amount was intended to cover 30 million starving people.

John Shore, later to be Lord Teignmouth and Governor-

General of India, first arrived in Calcutta as a civilian during the middle of the famine. He was moved to write some forty years later:

> Still fresh in memory's eye the scene I view,
> The shrivelled limbs, sunk eyes, and lifeless hue,
> Still hear the mother's shrieks and infant moans,
> Cries of despair and agonising groans,
> In wild confusion dead and dying lie:
> Hark to the jackal's yell and vulture's cry,
> The dogs fell howl, as, midst the glare of day,
> They riot, unmolested, on their prey,
> Dire scenes of horror! which no pen can trace.
> Nor rolling years from memory's page efface.[25]

All too late, the East India Company woke up to the crisis. 'The streets were blocked up with the promiscuous heaps of the dying and dead,'[26] described one letter back to the Company's Court of Directors in England. 'Interment could not do its work quick enough, even the dogs and jackals, the public scavengers of the east, became unable to accomplish their revolting work, and the multitude of mangled and festering corpses at length threatened the existence of the citizens.'[27]

When the rains finally arrived in the summer monsoon of 1770, and crops eventually did sprout, many of the villagers were no longer there to harvest them. Ten million had died, including a third of the population of Bengal, and hundreds of villages were left totally uninhabited.

Even the aristocracy was hit hard. Over two-thirds of the *zamindars*, or hereditary landowners, became bankrupt after finding it impossible to collect rent from the peasants. Most were forced to sell to entrepreneurs from Calcutta, breaking the link between landowner and tenant and ushering in a new era of absentee landlords. The loss in revenue, along with the first serious depopulation of Bengal in centuries, led to a drastic decrease in Company profits, and a financial crisis in 1772. The Court of Directors finally admitted that revenue collection had been 'equally impolitic in its institution and oppressive in the mode of exaction'.[28] New laws were passed in Britain to increase state interference in the Company's affairs, paving the way for full British rule. However, the resentment left in the minds of Indians towards the British did not fade quickly.

1844–6[29]

In the early summer of 1845, a consignment of potatoes from the Americas docked in Belgium. Intended as a new source of seed potatoes to counter the scourge of potato dry-rot, the shipment only succeeded in replacing a minor problem with an infinitely greater one. The fungus *Phytophera infestans*, more commonly known as potato blight, had attacked crops across North America for years, but Europe's unfavourable climate had prevented it taking hold across the Atlantic. The weather, in the early summer of 1845, was totally out of character in Europe.

When conditions suit it, the fungus will spread through a plant's vascular system, turning the roots and tubers into brown slime. In order to proliferate, high humidity and a high minimum temperature are needed for the disease-carrying spores to form. The European summer of 1845 was ideal for this process. Some areas in the north, such as Scotland, Norway and Poland were too cold, and much of the east was too dry. But across large parts of Europe, spores spread out like rippling waves from their epicentre in western Flanders. With humidity over 90 per cent for nearly ten weeks on end, free water could form on the potato's leaves, allowing the spores to germinate. The crop toll was high, with regions such as Belgium and the Netherlands losing up to three-quarters of their harvest. In Portugal, failure of the potato crop occurred at the same time as the government introduced changes to land registration, which in effect disadvantaged the illiterate peasantry. In 1846, a rebellion, led chiefly by women fearing a loss of their land rights, broke out in the north of the country, ushering in the third and last phase of the protracted Portuguese Revolution.[30]

The unusual summer weather in Europe had not gone unnoticed. In fact, the general consensus was that 'the highly charged state of the atmosphere'[31] was the direct cause of the loss of potatoes. Except for that of a few botanists, no causal link was made to the rusty brown spots on the leaves. In the autumn of 1845, farmers were unaware of the dangers of leaving the diseased fields untreated or of discarding the infected potatoes into furrows. Unfortunately, the winter was exceptionally mild, and the potato fields acted as incubators for the infection. Planting of the new crop began late because of continuous spring rain, only to be followed in May and June by an unprecedented heatwave, retarding potato growth. In most of Europe, the main cause of concern was drought. In Ireland and the west coast of Scotland, a series of Atlantic depressions that were blocked by persistent high pressure over central Europe kept Irish skies overcast and even more humid than in the previous year. The time-bomb in the potato fields exploded. And whereas other communities in Europe had survived the previous year's potato failure by turning to other crops, in Ireland the potato was the people's lifeblood.

Much of the Irish land belonged to alien landlords, and a large proportion of the remainder was divided into tiny plots that fed the rural population. From such small areas, potatoes provided large returns, especially from the poor-quality Lumper variety, a strain particularly susceptible to potato blight.

In some people's eyes, the reliance on the potato was not

necessarily a bad thing. The social philosopher Adam Smith said of the potato:

> The strongest men and the most beautiful women perhaps, in the British dominions, are said to be the greater part of them from the lowest rank of people in Ireland, who are generally fed with this root. No food can afford a more decisive proof of its nourishing quality, or of its being particularly suitable to the health of the human condition.[32]

Smith's admiration for the potato may not have been widely known in England, but it was the British government's over-zealous interpretation of his economic principles that con-tributed so much to the human tragedy in Ireland, just as it did later during the 1887–8 El Niño in India. Even though Smith never ruled out public relief, his principles were used to justify a lack of state intervention. What he did oppose was any suppression of free trade. Despite the burgeoning famine, instructions were issued not to interfere with the private market, thereby encouraging food to flow into the country according to demand. Demand required people to have money, and without it little food arrived.

Even when the weather improved in 1847, allowing pota-toes to be grown for the first time and hindering development of the blight, there were few seed potatoes left to be planted. By then, disease and fever were rife. By 1851, the population of the island had plummeted by nearly two million, of which death through starvation accounted for roughly half and emigration the remainder.

1888[33]

In 1888, El Niño combined with another microscopic organ-ism, this time killing a third of Ethiopia's population and changing the course of the country's history. The event is still known today as *Kifu Qan*, the Evil Days.

At the time, life in Ethiopia had changed little since the Middle Ages. There was no market economy, the land was run by feudal lords, and much of the population had been impoverished by a century of intermittent wars. The country was ruled by Emperor Yohannes IV, while the province of Shoa in the south was controlled by the ambitious King Menilek II. Along the coast, the Italians had established several colonies, with their eyes fixed firmly on the prize of a much larger chunk of Africa.

Then, as now, the Ethiopian economy was based mainly on rain-fed agriculture. In 1888, both the short and long rainy seasons had failed, and by the end of the year all the crops had shrivelled in the sun.

In the countryside, grain rose to over 100 times its normal price. Cow skins were cut up into pieces and ground into powder to make cakes. Children looked for grain in the dung of camels. Christians sold themselves to Muslim traders, and cannibalism was rife. From a popular song from Shoa came the line 'his wife gave him indigestion',[34] after one man was caught eating his family. Subtly, the balance of nature had changed. People followed the flight of vultures to locate carcasses. Scavenging animals abandoned human corpses in favour of the living. Even in the streets of Addis Ababa, cries rang out at night as people were dragged away by hyenas and jackals. The country, as typified by a poem at the time, was enveloped in a sense of utter hopelessness:

> When I try to flee from the poverty,
> It follows me,
> It perches on my head and sings and dances.[35]

The switch from El Niño to La Niña in 1889 failed to bolster either of the two rainy seasons in Ethiopia. By that time, another problem had arrived to exacerbate the misery. Two years previously, a cargo of imported cattle landed at the port of Massawa for the expanding Italian colony. With the shipment came rinderpest, a highly contagious virus related to measles and distemper, and the most severe infectious disease in cattle. The contagion spread from Massawa like wildfire, killing up to 90 per cent of the stock and even decimating herds of wild buffalo and antelope. Although the fundamental problem was a lack of rain, the arrival of rinderpest left few cattle to plough the fields, or people with the strength to use a hoe.

The continuing famine presented just the opportunity the Italian colonists had been waiting for. Not wanting to lose out in the European scramble for Africa, the Italians had no intention of remaining cocooned in Ethiopia's coastal lowlands. Their envoy Antonelli witnessed the impact first-hand. An area he knew to possess 'an atmosphere of abundance and prosperity' now struck him as 'one continuous desolation'.[36] In February 1889, he advised his government to seize the opportunity. 'The occupation of places in order to trace a frontier could be carried out this year in the certainty that we will not have great difficulties to overcome'. Although the invaders found it difficult to march into zones totally stripped of supplies, the sight of abandoned lands gave them the moral justification they sought for their policy of resettlement.

However, someone else was also making the most of the famine to advance his own plans for expansion. In the south, King Menilek used the lack of potential resistance from the starving semi-nomadic tribes to continue his takeover of several kingdoms and states, securing the boundaries for Ethiopia that are largely intact today. And when Emperor Yohannes died fighting the Sudanese Mahdi in 1889, Menilek was the strongest contender for the throne.

As emperor, it was important for Menilek to show

sympathy for his people's plight. Typifying the grand, symbolic gestures he later became known for, Menilek refused to eat meat, and was even seen toiling the soil with a hoe. He excused a woman who had confessed to eating seven of her children, and paid for her remaining son to be sent to school. Yet for all his outward benevolence, Menilek's desire above all others to maintain his army was extracting a very heavy price in the countryside. With so little grain available, Menilek introduced a new tax on the peasantry, one that his chronicler described disingenuously as 'popular'. When the revenue became difficult to extract, the armies raided any hidden stores. The tax and the ruthless methods used by soldiers to collect it resulted in the further deaths of hundreds of thousands of people already weakened by famine and rinderpest. But in spite of plundering whatever they could, the armies still failed to gather sufficient supplies or horses to stop the Italian occupation of Eritrea.

With over 90 per cent of the country's cattle lost, and continuing poor and erratic rains, the famine dragged on into 1891, when the next El Niño rubbed salt into the country's festering wounds. Only with the arrival of a La Niña in mid-1892, bringing good summer rains and the first decent crops for five years, was the country put back on the slow road to recovery.

The famine marked a turning point in the history of Ethiopia. Although at an appalling cost to the people of the countryside, it cemented Menilek's position and enabled his armies, when stronger, to defeat the Italians in the Battle of Adowa and repel their advancement. In so doing, Menilek spared his Ethiopia from European colonisation, the only African country to escape such a fate. The other winners were the Catholic and Protestant missionaries who succeeded in attracting many 'infidels' with offers of rice in exchange for conversion. A much larger country emerged from the famine with similar boundaries to those of today. Its leader, who had used the drought to consolidate his power, would go on to achieve great steps towards the modernisation of Ethiopia and become fêted as one of its greatest ever leaders.

At the same time in Russia, a long period of economic stagnation and gross social inequality helped turn its El Niño induced drought into a famine of profound national importance.

1891–2[37]

The 30 million or so peasants of Russia's Volga region were also cursed with the abnormal weather of 1891. A harsh winter had affected crops all over Europe, and was followed by spring frosts just as the seeds planted the previous autumn began to germinate. Farmers were desperate for a decent summer harvest. What they got was an exceptionally dry three months, emptying the wells and burning most of the crops that remained.

Once the animals had been sold for grain, and the grain had been eaten, a 'bread' made of rye husks and moss was all that was left for many. The misery was compounded by an outbreak of dysentery and diarrhoea, adding to the crisis because human dung was needed as the main fuel for heating.

At first, the appalling state of the roads and communications, along with the Russian tendency towards secrecy, meant few people outside of the region knew of the human catastrophe that was developing within. The official response was to deny the famine. Cheap grain continued to be exported in response to the Europe-wide poor harvest, and discussion of the famine in the press was discouraged. There was one man, however, whom the state could not, or more pertinently dared not, suppress. By 1891, Leo Tolstoy was in his sixties, his international reputation as an author and moral philosopher having made him one of the most revered men in the world. It was a matter of acute political embarrassment, as well as great significance to the outcome of the famine, when he stopped writing.

'It is more important to love than to feed,' was Tolstoy's initial reaction, believing that 'the most effective remedy against famine is to write something which might touch the hearts of the rich.'[38] But once he became aware of the extent of the crisis, he quickly changed his stance and played a personal role in relieving the suffering. For two years, Tolstoy and his family ran a series of soup kitchens, feeding up to 13,000 people daily. Despite hostility from the Church and a lack of official sanction from the government, his efforts at publicising the famine internationally attracted enormous financial support, as well as pricking the conscience of his own country's liberal classes.

Tolstoy was not the only literary figure inspired. Anton Chekhov, a trained physician not yet widely known as a writer, was moved to a similar sacrifice:

> There is no time to write. I abandoned literature long ago, and I'm poor and broke because I thought it desirable for myself and my independence to refuse the remuneration cholera doctors receive . . . The peasants are crude, unsanitary and mistrustful, but the thought that our labours will not be in vain makes it all unnoticeable.[39]

By the time a successful harvest was reaped in 1893, over 650,000 people – and by some accounts many more – had died from starvation, cholera and typhoid fever.

Revolution in Russia did not occur for another twenty-five years after the famine of 1891–2, yet the events of those years greatly weakened the authority of the Tsar. Encouraged by Tolstoy's attacks on social injustices and the Church, a new period of dissent blossomed that by

agreement of many historians became a milestone on the
road to revolution.

1930–1[40]

In 1930 and 1931, the French colonial administration displayed similar insensitivity to that of the Russian state when
adverse weather gripped Vietnam. Its indifference to the
plight of the local population resulted in a revolt against
French rule that foretold of a much bloodier fight for independence to follow.

In north and central Vietnam, following extensive flooding
the previous year, a severe drought extended throughout
1930 and beyond the spring crop season of 1931. At first,
when the rice crops failed, peasants turned to the cheaper
alternatives of millet and beans. But as famine took hold
many were forced to resort to the gathering of edible roots in
the forest. 'I had never seen such a sight as that at the *soupes
populaires* [soup kitchens],' wrote one doctor, 'thousands of
walking skeletons with absolutely nothing to eat, true cada-
vers whose ribs jut out under the skin.'[41] It was an appalling
and, as it turned out, most misjudged point for the administration to raise its hated 'head-tax' in 1931. One missionary
wrote of the decision: 'Since 1930 three harvests have been
lost and the fourth, that of the tenth month, appears in
jeopardy. The collection of taxes was abruptly made at the
moment when the population was ruined by drought. In order
to pay, the inhabitants have had to sell off all their property.'[42]

Following agitation by the Communist Party, protests and
civil unrest broke out across the countryside, and 6,000
peasants marched on the city of Vinh. The French administration was driven out of the provinces of Nghe An and Ha
Tinh where large estates were shared among the population,

and hundreds of autonomous 'soviets', or village republics, established. The scale of the harassment and killing of colonial officials was so great that the year 1930 became known as *La Terreur Rouge*, or the Red Terror.

The following year, when the French re-established control, became even more infamously known as *La Terreur Blanche*. Reprisals against the local people cost 10,000 lives and another 50,000 deportations. At the end of the year, the Governor-General felt confident enough to declare that 'communism has disappeared'.[43] The party's leadership was certainly decimated, but the period only served to consolidate its powerbase, and it re-emerged even stronger during the Second World War.

1972–3[44]

In the story of the unfolding scientific understanding of El Niño, the events that took place in 1972–3 were a major turning point. Occurring only a few years after meteorologist Jacob Bjerknes published his seminal paper on its physical processes, it was really the first El Niño to grab the attention of the scientific world, largely because of the widespread political and economic havoc that it wreaked.

For a start, the extraordinary collection of climatic anomalies had a profound effect on global grain prices. Cereal output in Russia was only 90 per cent of that of the previous year, after a dry summer in the west ruined crops. The Russian government was forced to suffer the ignominy of importing six million tons of grain from its American archenemy. India and China were also forced to meet drastic shortfalls in their harvests with imports. Ordinarily, America's grain producers would have responded to the demand in the global grain market by planting more. Ironically, El Niño's impact elsewhere presented them with an even more lucrative option.

In Peru, the increase in coastal sea temperatures characteristic of every El Niño had triggered the collapse of the
country's anchoveta fishery. Anchoveta is the source of fishmeal, the favoured protein supplement for the North American poultry industry, a market so important that by 1972 it
had made Peru's anchoveta industry the biggest fishery in the
world. With anchoveta scarce, American farmers found that
soya bean, an alternative to fishmeal, fetched even higher
prices than wheat, despite the major grain crisis at the time.
Coupled with the revelation that the Russians had negotiated
large quantities of US wheat at subsidised low prices, dubbed
'The Great Grain Robbery' in the press, grain prices soared.
And for the first time in more than twenty years, global food
reserves fell, and politicians and economists were forced to
question their world's ability to feed its ever-growing population. Elsewhere, the crisis caused by the climatic anomalies of
that year resulted in the overthrow of at least three political
regimes.

In Afghanistan, a groundswell of general resentment built
up during the country's worst drought in recorded history.
Official figures stated a death toll of 80,000 – the most the
government was prepared to admit. Conditions in the countryside, in part a consequence of corruption and inept administration, were starkly contrasted by those enjoyed by the
élite, including King Zahir Shah. Pro-Soviet organisations
exploited the frustrations of the population, who were further
enraged when famine assistance was sought from American
and other Western relief organisations. The end of the King's
rule, and of the monarchy in Afghanistan, came in July 1973,
while the King was away in Italy for medical treatment. He
was deposed in a bloodless coup, and those leading the new
republic began an ultimately costly period of closer co-operation with the Soviet Union.

Sub-saharan Africa had been experiencing below average

rainfall since 1968. The global atmospheric changes of the 1972–3 El Niño, superimposed on the longer term cycle of low rainfall, worsened the situation in many countries in the region.

In Niger, President Hamani Diori paid the price for not doing enough. Throughout both years, the country saw less than half its average rainfall. The nomadic Tuareg, Bororo and Fulani tribes, representing a third of Niger's population, were the hardest hit, but crop failures in the southern agricultural belt extended the food shortage throughout the country. Added to the damaging impact of the international oil crisis, tax revenues fell by 40 per cent. Access to international relief supplies, distributed through a corrupt and insensitive administration, was tied to proof of tax payment, which most people were unable to pay. By 1974, death or migration had reduced the population by a quarter and over 80 per cent of the country's cattle had either perished or been driven south. The same year marked the end for President Diori, who was overthrown by a disaffected army while attending President Pompidou's funeral in Paris in April.

In Ethiopia, King Haile Selassie also paid the price for the lack of rainfall. But whereas Diori's response to the crisis was deemed insufficient, Haile Selassie may not have known about it at all.

For much of Ethiopia's northern highlands, particularly in the provinces of Tigre and Wallo, the harvests of both 1972 and 1973 produced less than half their expected yield. The population had trebled in the previous quarter of a century, and in a large proportion of the drought-stricken provinces, the land was in the hands of absentee landowners, who commonly demanded between 50 and 75 per cent of the tenant's produce as rent. By the end of 1973, over a million people were close to starvation and at least 200,000 had died, including a quarter of the nomadic Afar tribe.

The king was over eighty years old, living in a twilight zone of increasing senility and presiding over a feudal and stagnant bureaucracy. Selassie's chief concerns were his reputation as an African statesman and his prospects of winning a Nobel Peace Prize. District officers who came with tales of the unfolding famine in the north were rebuked for reporting an occurrence so frequent in the country's history. As recently as the 1965 El Niño drought, for instance, more than a quarter of a million people were said to have died in Tigre alone. 'Since this was eternal and normal, none of the dignitaries would dare to bother His Most Exalted Highness with the news that in such and such a province a given person had died of hunger,'[45] a court official related in Ryszard Kapuscinski's *The Emperor*.

With no official relief forthcoming, student demonstrations erupted on the streets of the capital. When stories in the foreign press exposed the severity of the famine that had been allowed to develop unnoticed, officials were embarrassed into accepting international support and mounting their own ad hoc relief programme. But it was too late. Junior army officers began agitating for change, and broadcast a pirated copy of a BBC documentary about the famine which they had intercut with images of the emperor. A huge outcry followed, and the army mutinied in the south. By September 1974, the mutiny had become a revolution. Haile Selassie's fifty-eight-year reign was over, ending a 2,000-year-old monarchy, and a new Marxist Mengistu government was ready to plunge Ethiopia into an even worse period of corruption and turmoil

Revolution and Conflict

Power is necessary, but is confined within precise limits. It requires consent and a certain reciprocity . . . In a more general way, it might be said that power must justify itself by main-

taining a state of collective security and prosperity. This is the
price to be paid by those who hold it – a price that is never
wholly paid.   *Georges Balandier, Sociologist*[46]

As we have seen, recourse to wars, rebellions and revolutions
has been a recurring response to some of history's severest El
Niño events. Famines and other natural disasters are quick to
undermine a community's confidence in its rulers and the
legitimacy of its religious order. In this way, they can act as
springboards to those waiting in the wings of power or spurs
for popular uprisings. In fact, in some countries, the link
between conflict and El Niño related drought has been a
repeated theme throughout history. The colonial pasts of
Mexico and South Africa are two examples.

Mexico's weather, although undoubtedly influenced by El
Niño and La Niña events, does not exhibit a particularly
consistent signal. In fact, Mexico straddles two separate
regions with opposite responses during El Niño events. On
the one hand the central America isthmus tends towards
drought in El Niños and towards heavier rainfall during
La Niñas, while northern Mexico and south-western USA
on the other hand lean towards the reverse. However,
despite El Niño's inconsistent effects on Mexican weather,
El Niño induced drought and floods have brought horrific
suffering to the population over the centuries. Since the
Spanish invasion in 1519, the consequences on many
occasions have had a momentous effect on the country's
history.

The first recorded incident happened in 1624, and is often
cited in Mexican history as an important precursor to even-
tual independence. The drought that year was the worst since
the Conquest, forcing maize prices to unprecedented heights.
When the Marquis de Gelves, the Spanish Viceroy in Mexico,

attempted to control local speculators, serious rioting erupted and forced him back to Spain.

The year 1692 became known for the Corn Riots. Following a year of floods, insect infestations and highly unusual April snowstorms throughout the Valley of Mexico, a shortened growing season drained the last of the grain stores. Rumours spread of the authorities' misuse and hoarding of grain, and mobs of starving Indians descended upon Mexico City, stoning officials and burning the viceregal palace to the ground.

The El Niño of 1785–6[47] brought similar scenes to the streets of the capital. The unusual weather began with a rainy season that finished almost as soon as it began, and was compounded by severe frosts in August, several months earlier than normal. Tensions, already brewing over the injustices of landownership and colonial rule, exploded when news spread of landowners and merchants harbouring dwindling stores to feed their own families and drive up prices. The granaries of many *haciendas* were burned and thousands of indigenous people fled to Mexico City, leaving the roadsides littered with the bodies of those who failed to make it. The period became known as 'The Year of Great Hunger', but few lessons were heeded by the Spanish authorities.

As every Mexican schoolchild knows, not long after midnight on 16 September 1810, the Creole priest Miguel Hidalgo y Costilla rang his church bell and called Mass early. His call for independence from Spain was enthusiastically received by the small congregation, who then set out for the next village in search of recruits to join the uprising. It was a completely spontaneous and disorganised protest, which could so easily have become an isolated act of treason, quickly suppressed by the authorities, and relegated to the dustbin of history. Yet Hidalgo had touched a nerve among thousands, and his call for independence sparked a full-scale insurrection against the Spanish Crown.

So how did a band of thirty people suddenly swell to 100,000? Records show that rainfall between 1808 and 1809 was erratic and low across much of Mexico. In the highland areas without irrigation, the summer rainy season was relied upon to produce the single annual crop. Trailing in the wake of two poor harvests, the last thing the country needed in 1810 was El Niño.

'Nowadays there is such need among the poor people that they cannot find anything to eat,' wrote the manager of one *hacienda* in August 1810 in a letter to his boss. 'The Indians from those villages are robbing cattle even in broad day-light.'[48] When two cattle rustlers were finally captured in the vicinity of the *hacienda* and taken to the local jail, it was already bursting with Indians forced into the same crime.

Hidalgo's War of Independence lasted only seven months, but resistance continued for a further ten years before independence from Spain was finally achieved. But independence did not prevent several decades of conflict and instability before Portirio Díaz won office in 1876. For more than three decades, Díaz ruled the country with an iron fist and while his tenure was characterised by economic stability and growth, virtually all the benefits flowed to the upper classes and foreign investors. Over 96 per cent of rural families owned no land at all, and 60 per cent of them were defined as 'debt slaves' – peasants indentured to work on *haciendas* because of an inability to pay off their loans. Small farmers too were already having to compete with the overwhelming advantages of the big *haciendas*, with their favourable taxes and peon labour. In times of drought, these disadvantages became even greater, because the small farmers' land was insufficiently irrigated and access to water was monopolised by the *haciendas*.

While the underlying reasons for the Mexican Revolution lay in the long-term economic condition of the peasants, its timing may also have been triggered by the weather. The year 1910 marked the culmination of three years of drought and the situation for many was becoming unbearable. 'The Indians have [left] their hovels as wolves leave their dens in winter and have set upon some *haciendas* in order to seize [whatever] grain they find,' a local periodical of the time recorded. '[What is more] about these assaults is that the Indians, although able to [steal] other valuables, have no interest in taking [anything except] maize, beans, etc., which shows that we are not dealing with robbery or [matters of] the criminal code, but with hunger.'[49]

The years 1908–10 were La Niña years, which as a generalisation tend to bring drought to the north of Mexico. It is perhaps no coincidence that the first revolutionary activity erupted in the northern states in 1910, where Francisco Madero's call for political change fell on fertile ground amidst the hordes of disgruntled peasants. Pancho Villa, Pascual Orozco and others leapt on the cause by mounting attacks on towns and *haciendas*. The final trigger for mass action came when Emiliano Zapata broke into a police station in March 1911, disarmed those on duty, and called a public meeting in the town square. Just as Hidalgo's actions had lit the fuse one hundred years earlier, Zapata's spontaneous action sparked thousands of peasants across Mexico to rise up against the government. By the end of the year, Díaz had abdicated.

There are no written records of drought in southern Africa before the arrival of Europeans in the seventeenth century, but it can be safely assumed that the region has always suffered periodic episodes of drought. Its seminal role in history, however, did not emerge until the colonial era. As white settlement expanded eastwards from the Cape, the constantly shifting boundaries between whites and blacks became the main interface for violence. With Africans being forced to give up their best pastoral land and access to water, suspicion and mistrust dominated relations. The appearance of drought every few years fuelled racial tensions on both sides and exposed the inadequacy of the Africans' remaining land to support its people in difficult times. Resentment erupted into conflict. Desperate Africans raided cattle and attacked homesteads, and the colonists responded with their own reprisals. One indication of the direct linkage between this catalogue of conflicts and drought is the fact that the majority of the nine Frontier Wars between the colonists and Xhosa were triggered during El Niño events.

By the 1820s, the combination of European expansion and drought was also beginning to impose a significant impact on the inter-tribal relationships of the Nguni, an ethnically distinct collection of chiefdoms from the south-east. The first three decades of the nineteenth century were characterised by low rainfall, with certain rainy seasons, particularly in the years of 1812, 1817 and 1821–3, providing virtually no rain at all. The driest period by far began around the turn of the century and lasted three years. It became known as *madlathule*, literally translating as 'let one eat what he can and say naught'.

It was a time, according to oral tradition, of famine, raiding, warfare and cannibalism. A radical political transformation swept the landscape which saw smaller and weaker chiefdoms consumed into larger ones that offered greater protection and security of food supply. The three powerful chiefdoms that grew out of the process continued to compete with one another for dwindling resources. Further disruption between the communities was fuelled by the European slave and ivory trade that developed from around 1820 from its base at Delagoa Bay. The Zulus, using superior weaponry and fighting tactics developed by their legendary leader Shaka

Zulu, emerged from this chaos as the single dominant kingdom. Their military superiority forced whole tribes to flee
their traditional homes in Natal, some in turn making their
own conquests further afield. This mass migration of people
led to the formation of Swaziland and Lesotho and sent some
refugees as far north as Tanzania and Zambia. In local
tradition, the time became known as the *mfecane* or *difaqane*,
or literally 'the crushing' or 'scattering'. At the end of the
period the tribal boundaries of vast swathes of southern,
central and eastern Africa had been irrevocably changed.

By the 1830s, a trickle of Boer settlers from the Eastern
Cape, disgruntled at the perceived inequities of British laws
and administration, began crossing the Orange River. With
parts of Natal, Transvaal and the Orange Free State left
severely depopulated, pioneering white settlers felt justified
in expropriating the land, and many Africans that were left
were prepared to accept mastery from white traders offering
food and protection. News of the 'vacant' lands filtered back
to the Eastern Cape, and the pressure to move mounted. But
when a series of droughts in the mid-1830s struck the Eastern
Cape, and in particular the drought of 1837–8, the trickle
became a flood, as thousands of settlers joined the seminal
movement in Afrikaner history, the Great Trek. Within a few
years, however, serious conflicts erupted with the indigenous
Africans. Many refugees had returned to their homeland only
to clash with the Boers who had staked their claims there.

When drought struck again in the major El Niño of 1844–
6, the plight of the Xhosa was desperate. Most on the border-
land to the Cape Colony and Ciskei had been crowded into
marginal land that was scarcely able to sustain them even in
good years, forcing many to squat on settlers' land or work as
labourers. Thieving of Boer cattle increased after the failure of
the Xhosa's crops, adding to the voices already calling for
them to be expelled from the Cape Colony. 'Thieves became

more and more daring and the chiefs less active to prevent them,' wrote one missionary. 'They [the Boers] conceived a war preferable to existing circumstances and were most urgent for the government to strike a blow, which it . . . most imprudently did.'[50] The spark was ignited when an Xhosa chief refused to hand back a prisoner who had been successfully rescued by a party of his men, after being held by Boers for stealing an axe.

The Seventh Frontier War, or the 'War of the Axe' as it more commonly became known, began in April 1846 and was ill-conceived and unplanned. The Xhosa were very reluctant to enter into a conflict during drought, understanding the deprivations that would ensue from fighting in such conditions. The British on the other hand seemed oblivious to the appalling timing. Soon, their oxen were dying and wagons carrying arms and equipment could not be pulled. The lack of food available to both people and livestock immobilised the troops – a situation exacerbated by the Xhosa deliberately burning grasses as they retreated. The Xhosa had the skills and experience to survive these periodic droughts, but they did not push for victory. No doubt they knew they faced the superior resources of the British who had access to fresh food supplies from the Western Cape, a region that during an El Niño typically receives higher rainfall than normal.

Without ever surrendering, the Xhosa stopped fighting in September 1846. It was the last month in which a reasonable harvest could be expected, and they dared not risk another crop failure and the months of food shortages that would follow. The British read this cessation of hostilities as a victory, and annexed the new land of British Kaffraria – or Ciskei as it later became known – to the Cape Colony.

The last and most decisive battles in the century-long series of black versus white conflict occurred in the late 1870s against the backdrop of the strongest El Niño of the century.

The stage for these wars had been set over a number of years. The British wished to amalgamate all white-ruled states into a confederation, and partly in order to protect their boundaries, needed to dismantle all neighbouring African kingdoms. Following the recent discoveries of major mineral deposits, they were also desperate for the tax revenue, new land and cheap labour that non-independent African states would provide. Military victories seemed the most feasible means to acquire them. With the arrival of drought in 1877, Boer insecurities over their border with the Zulus and claims for more productive land created considerable pressure on the British to act. Following the annexation of the near-bankrupt state of Transvaal to the British, demands by increasingly disgruntled Boers needed to be satisfied to ensure a stable confederation.

Recurring droughts had left the Africans acutely aware of just how much fertile land and access to dependable water sources they had lost. Many of the young warriors were agitating for decisive action against the whites, but most chiefs realised that the odds were stacked overwhelmingly against them, despite being better armed than ever before. The call of Sarili, chief of the Xhosa nation, to his councillors one night summed up his resolution to face the predicament. 'I intend to fight. I am in a corner. The country is too small, and I may as well die as be pushed into a corner.'[50]

The first battle took place in Eastern Transvaal in 1876 against the Pedis, who resisted fiercely for three years before succumbing to defeat. The Ninth Frontier War against the Xhosa followed in 1877 and lasted a year. But in order to achieve a confederation, the British had first to destroy the threat of the Zulus. The circumstances were engineered by the British Commissioner, and by late 1878 an Anglo-Zulu war was virtually inevitable. In December, seeking reparation for a minor Zulu incursion, the Commissioner issued an ultimatum for the full demobilisation and abandonment of the Zulu military system and a hefty fine in cattle. The Zulus were given twenty days to comply. The Zulu king begged for more time. The drought had finally broken, and the heavy rains accompanying the start of the 1879 La Niña allowed crops to be planted. The king knew the harvest would be delayed until the end of February at the earliest and his army was depending on it. In a bid to buy time, he told the British the rivers were in flood and it would be impossible to collect the cattle before the British deadline expired. But the British knew the state of the Zulu food supplies and guessed the offer was a ploy to stall until harvest. The requested extension was refused, and war commenced immediately the deadline expired. The British were overwhelmed by Zulus in the initial battle, but, swelled by thousands of reinforcements, defeated the Zulu army some eight months later.

Slavery and Migration

In spite of the many historical examples of uprisings and political overthrows in the wake of catastrophic weather, the élite have seldom emerged as the ultimate losers. Almost invariably, the poor have felt the full brunt of natural disasters and finished up too weak or disorganised to revolt. In most cases, climatic hiccups have only served to strengthen the hand of the landowners, slave traders, moneylenders and others keen to exploit the poor. Famines, floods and other catastrophes have therefore tended to widen the gulf between rich and poor and further undermine social equality. Faced with starvation, many poor people have been forced to flee their homelands in times of stress, most never to return.

In some regions, an intermittent pulse of people has poured out over the years in response to climatic crises. For most of these people there has simply been no alternative but to seek

relief and asylum elsewhere. This pattern of movement driven
by El Niño has led to large-scale, permanent demographic
change and in effect created diasporas of 'El Niño's children'.

Records of exceptionally strong events in historically recent
times show that El Niños have probably generated simultaneous waves of migration around the world. During the
1877–8 event, for instance, 400,000 people headed out of
Brazil's drought-stricken north-east, many opting to join the
Amazonian rubber boom. In India, 200,000 labourers from
the south migrated into Ceylon, while in the north of the
country, unknown tens of thousands moved into the low-
lying *teria* swamplands along the Himalayan foothills where
huge numbers succumbed to malaria. In China, the previously
uninhabited Changbai mountains of the far north-east were
settled by thousands of starving peasants escaping drought-
ridden provinces such as Shandong further south. And across
Indonesia there was widespread movement between islands,
with many people emigrating as far as the Malay Peninsula. In
most cases migration was permanent.

In earlier times, the mass movement of people was often
linked to the practice of slavery. The market for slaves acted
as a pressure valve for drought-afflicted zones. For the land-
less and dispossessed, enslavement was often a voluntary
choice to escape starvation. The first offered into slavery
were generally the most unproductive members of a family
– the children. In his vivid chronicle of the great Egyptian
famine of 1200–1, Abd al-Latīf recorded the pleas of a mother
to buy her 'not yet nubile' daughter. When he rejected her
offer, the desperate mother replied, 'Oh well then, take her as
a gift.'[52]

Before the advent of the international slave trade, slaves
usually wound up with chiefs and landowners. These arrangements were not necessarily permanent or inhumane. In some
parts of the world, landowners even regarded the adoption of

slaves in times of crisis as a paternalistic duty. The common practice of slavery did, however, perpetuate the economic conditions that left the poor unable to cope with any downturn in food supply. In the seventeenth century, the role of climate in demographic change escalated with the start of a much more sinister form of slavery – an international trade driven by profit. Slaves once retained in or near their homeland were now taken much further away. Many were never able to find their way back.

Droughts often altered the economic relationships between regions. In some areas, droughts were even welcomed for the pulse of business activity they brought with them. The Nunu fishermen of the middle Zaire River, for example, needed labour to construct dams and levees, which they could normally procure by buying witches, adulterers and debtors from neighbouring agricultural communities. In times of drought, however, which did not affect the fishermen, the Nunus' purchasing power increased and they could even buy wives in exchange for fish.[53]

On the Coromandel coast of southern India, the change in trade was on a much more systematic scale, as described by the sixteenth-century traveller Duarte Barbosa: 'In some years it so happens that no rain falls, and there is such a dearth among them that many die of hunger, and for this reason they sell their children for four or five fanams each. At such seasons the Malabares bring them great stores of rice and coco-nuts and take away ship-loads of slaves.'[54]

Slave traders soon learned that drought was good for business. As a result, they eagerly sought out news of drought along their trading routes to cash in on the plentiful supplies of cheap slaves. It is no coincidence therefore that drought years often correlated with those years in which the greatest numbers of slaves were traded. Portuguese records from the west coast of Africa show a direct linkage between the extent of drought and warfare, two factors in themselves that are often inter-connected, and the numbers of slaves transported. During the drought of the great 1877–8 El Niño, for example, the licensed slave recruiters, Banco Nacional Ultramarino of Angola, recorded a record cargo of 512 slaves on a ship bound for the Sao Tome sugar plantations. This compared with the company's average load of between 50 and 200 slaves. But while trade in times of drought was brisk, it also had its downside. Many of the slaves bought or captured during these periods were already very weak and far less likely to survive the gruelling journey to their final destination.

In Brazil, drought in the country's poverty-stricken north-east was often the driving force behind waves of migration to booming areas elsewhere in the country. In the eighteenth and nineteenth centuries, much of this movement involved slaves who had become idle mouths for their owners to feed while waiting for a drought to end. In the worst droughts, such as 1877–8, landowners found themselves losing their cattle, cotton and even seed, and slaves became 'the only money in circulation'.[55] The eighteenth-century booms in gold, emeralds and diamonds were built on the backs of the north-east's *flagellados*, the flagellated ones, as they described themselves. By the nineteenth century, the coffee plantations in the south had become the main destinations for slaves. In the first twenty-five years of the century alone, 220,000 people left the drought-prone north-east, of which 80 per cent were slaves.

Slavery remained legal in Brazil long after all other countries, but the drought of 1877–8 marked the beginning of the end. The north-eastern state of Ceará suffered appallingly, with half of its population succumbing to starvation and disease. The particularly poor treatment of slaves in the north-east, highlighted by scenes of them amassed and emaciated on the beaches awaiting transport to the coffee industry, helped spawn a popular abolition movement in Ceará. In 1884, Ceará outlawed slavery – the first state to do so. From that point, pressure on slave-owners and the government mounted until emancipation for all was finally achieved in 1888, just as the next great drought was taking grip. But even the end of slavery did not halt the mass impetus to leave the north-east in times of drought. Huge waves of people moved to the Amazon in 1888, 1915 and 1931, to the south and central parts of the country in 1942, and to the newly built capital of Brasilia in 1958.

The international end of slavery that began with the abolition of the slave trade in Denmark in 1802 and was followed by the emancipation of slaves by Britain in 1838 did not end the ruthless exploitation of foreign workers. Many writers have argued that even where it was outlawed, slavery was disguised in various forms of labour schemes for several more decades. However 'free' these labour schemes purported to be, the fact is that as overseas markets emerged, and the need for cheap migrant labour increased, El Niño's visitations continued to drive strong pulses of migration right into the twentieth century.

In India, migration to cities and neighbouring states has always been a common response to the severest cases of monsoon failure. In the nineteenth century, many of these refugees went overseas. The rigid policy of free trade enforced by the British administration ruled out public relief schemes, and for many migration was the only alternative to starvation. With the vacuum left by the abolition of slavery, colonial planters were demanding cheap labour for their sugar, cotton and tobacco plantations throughout the British Empire. Labour indenture schemes provided the solution. Workers drawn mainly from the poorest states, such as Bihar, eastern Uttar Pradesh and Tamil Nadu were required to work for a fixed term – usually five years – during which time they were treated as virtual slaves. Some returned home at the end of their contracts, while others stayed behind to establish communities that still exist today.

Mauritius was the first important market for Indian labour. In the early 1840s, the prohibition of imported labour was lifted to service the sugar plantation boom. The relaxation coincided with the 1844–6 El Niño that helped to generate large numbers of migrant workers, particularly from the Tamil districts of the south. Many of these workers, and those that followed them, remained to form a sizeable Indian community in Mauritius that still exists today. Similar stories unfolded in other parts of the British Empire. Markets for cheap Indian labour developed in the Caribbean, Natal, Fiji, Guyana, Malaysia and Burma where significant Indian communities also formed. The numbers migrating were always high in times of famine. During the El Niño of 1877–8 more than 300,000 people left Madras alone, most of them destined for Ceylon. In times of drought, traders had little trouble in recruiting willing labour: 'In most cases the recruiter finds the coolie absolutely on the brink of starvation and he takes him in and feeds him and explains to him his service,' wrote one emigration agent during the drought of 1877, and '. . . under such conditions, our terms of service are absolute wealth'.[56] Another agent in India, struggling to find recruits to go to Fiji during the La Niña of 1904, witnessed the flipside of this coin. 'The recent harvest in India has been exceptionally good, and the result is that emigrants are at present almost unobtainable.'

In the second half of the nineteenth century, the labour trade worked hand-in-hand with drought throughout the tropical Pacific, depopulating several of its islands. On some islands, this combination was arguably an even greater agent for change than the arrival of the missionaries. Ironically it was the abolition of slavery in America after the Civil War that precipitated the devastation of Pacific communities. By depriving the southern plantations of their labour, abolition ended American cotton exports and helped spawn new plantations in Queensland, Hawaii, Fiji, Tahiti and Samoa to meet the international demand. These plantations in turn sought cheap labour, and along with other burgeoning industries such as sugarcane, coconuts, mining and coffee, found the Pacific islands the most fertile sources of migrant workers, particularly during droughts.

The official term for the trade was 'labour recruiting', but the more common name by which it was known, 'blackbirding', betrayed the racist overtones behind the recruitment practices and subsequent treatment of workers. In theory, the Pacific islanders were contracted of their own free will, but in reality they were often kidnapped or tricked onto boats, and the conditions that many later faced on the plantations bordered on slavery.

News of drought drew blackbirding ships like magnets. At these times, even the wary could be lured on board with promises of plenty to eat. According to John Bradley, a labour agent on the Gilbert Islands, potential recruits put up little resistance. 'Wages and terms of service were nothing to them. Food and water was all they craved. They would have sold themselves for life, for these two simples.'[57] But serious cases of drought brought another challenge to labour recruiters – locating those fit enough to work. Bradley also reported that many '. . . were so reduced that they would be no good as labourers'. Some of the labourers returned home, some forged new homes on the other side of the Pacific and many others died on the blackbirding ships or on the plantations. In all, less than 20 per cent ever saw their islands again.

The smallest islands faced the greatest risk of drought. Their limited freshwater supplies and scant productive land afforded little refuge or alternative food sources in times of

need. The isolated island of Banaba, or Ocean Island, which
sits virtually astride the equator 400 kilometres west of the
nearest Gilbert Islands, is a classic example. Here, there are no
natural pools of surface water, and any rain soaks immediately through the porous coral, accumulating in underground
caves, or *banga-bangas*. These pools were traditionally so
treasured that only selected women were allowed to descend
the long underground tunnels, blackened by hundreds of
years of torch-flame, to collect water with their coconut shells.
In times of drought, the only alternative for locals was to suck
the watery eyes of fish for moisture, or to follow rain showers
way out to sea in their canoes. The worst crisis for the island
occurred in the early 1870s, when the *banga-bangas* completely dried up during an extended La Niña event. Of those
that had not already succumbed to starvation, three-quarters
'opted' to board blackbirding boats to Hawaii. In 1873, one
boat bound for Queensland was loaded with people so weak
they could barely stand. Another group of Banaban workers
returning from Tahiti took one look at the state of their island
and opted to re-sign with the blackbirders. Before the
drought, the population had been estimated at between
two and three thousand. By 1880, it was down to thirty-
five.[58]

5

Uncharted Territory
Exploration, Settlement and Shattered Reputations

But, although it [the trade winds] blows from the East almost continually in those latitudes, we were favoured, during our voyage of 800 miles, with a fair wind, which was so light as to appear almost sensible that it was filling sails which could not endure its fury, while the sea was so smooth that it seemed as if reserving its power for some bark better fitted to withstand it . . .

The Reverend John Williams, missionary,
sailing Rarotonga to Tahiti on
'The Messenger of Peace' March 1828[1]

If El Niño's periodic occurrence has demonstrated great potential for human disaster, in the history of exploration and settlement it has served as both a hindrance and a help. Its unpredictable appearance has not only foiled and confused explorers, but in some cases it has created unique opportunities to reach and colonise new lands. A brief dip into the mixed bag of history turns up an assortment of fortunes.

When Europeans first discovered Polynesia they were astonished at the local people's understanding of the winds, rains and seas in all their variations. But given their primitive canoes and lack of navigation equipment, the Polynesians' incredible journey almost a quarter of the way around the globe to colonise the Pacific is one of the great mysteries of exploration. Even the great navigator Captain Cook was amazed. 'How shall we account for this Nation having spread itself, in so many detached islands, so widely disjointed from each other in every quarter of the Pacific?'[2] he asked in his journal. This extraordinary feat remains a puzzle to this day. Could the answer lie in the Polynesians' knowledge of El Niño?

Linguistic and archaeological evidence has now conclusively proved that the Polynesian ancestors came from the Asian side of the Pacific, but the mystery that has long puzzled geographers was how such a migration could be achieved against the direction of the prevailing south-east trade winds and currents. So strongly did Thor Heyerdahl believe that such migration was impossible, and primitive man could only ever travel in the direction of the prevailing winds and currents, that it underpinned his theory of an east–west migration, and subsequent balsa-raft voyage in the *Kon-Tiki*.[3]

The problem is that while sailing into the wind is possible, it is simply not feasible. Based on his voyages in a replica canoe, archaeologist Ben Finney has estimated that such a vessel could consistently sail at no better than 75 degrees off true wind.[4] At such an angle, a Polynesian canoe would have to tack 3.9 miles to gain one nautical mile in true distance. To cover a 400-mile journey, such as between Vanuatu and Fiji, in average conditions of a six knot sailing rate against the trade winds and a one knot current, a Polynesian canoe would need to cover over 4,300 miles of open ocean. With a vessel laden with the people, supplies and breeding stock necessary to begin a new colony adding even greater difficulties, the likelihood of such a journey against the trades would have been highly improbable.

The common description of the trade winds as prevailing means they are predominant, not, as Heyerdahl formerly believed them to be, 'permanent'. Each southern hemisphere summer, the trade winds weaken and, particularly west of the Dateline, are often replaced by monsoonal westerlies. Such winds usually come in short bursts of a few days to a fortnight at a time – long enough for the Polynesian mariners to bridge short distances. Not surprisingly the first stage of colonisation down through Melanesia into western Polynesia, characterised by relatively short distances between archipelagos, was completed within the relatively short period of a few hundred years, beginning around 1500 BC.

The second stage of colonisation into eastern Polynesia began after a 'mysterious pause' of a millennium or so. The factors that had assisted such rapid movement before were no longer so helpful. The annual westerlies that favoured colonisation of western Polynesia become weaker, less reliable and potentially stormier, the further east one ventures. In addition, the distance between archipelagos is often greater, particularly to the far-flung islands at Polynesia's extremities, such as Hawaii, New Zealand and Easter Island.

Reaching the Marquesas, the first island chain in eastern Polynesia to be settled, necessitated a journey of at least 1,800 miles. And if the Marquesan ancestors came from Fiji, as pottery evidence suggests, the journey would have been over 2,200 miles.[5] Likewise, settlement of Easter Island, most probably originated by Marquesan islanders, involved a journey of over 1,900 miles.

There may have been much longer journeys from the margins of Polynesia eastward to the southern and central American coast. No such journeys have been proven, but various scholars have uncovered tantalising clues to suggest that prehistoric trans-Pacific journeys were achieved. For instance, the Spanish *conquistador* Vasco Nuñez de Balboa stumbled across a strange group of black, frizzy haired men as he traversed westwards across Panama.[6] The unfamiliar-looking individuals were being held prisoner by local Indians, part of a larger group enslaved on the Pacific coast. At the time, the alien men were presumed to be of African origin, possibly because Polynesians were unknown at the time. Other clues of possible Pacific journeys, evident in such items as coconuts, bananas, domestic fowl, ceramic and architectural styles, all likely to be of Asian origin, were also discovered in the Americas by the earliest European visitors.[7] The South American sweet potato was already being cultivated in Polynesia when Europeans first arrived, a possible example of two-way trade.

Finney was one of many people puzzled by these journeys. In the 1970s, he had sailed between Hawaii and Tahiti in the *Hokule'a*, a sixty–foot, double-hulled reconstruction of a Polynesian canoe, thus proving long-distance voyaging in such vessels was possible. But the mystery of how to make such a journey against the trade winds remained. He was aware of El Niño's impact on Pacific winds, but the limited data of the 1970s suggested their effect did not extend eastwards beyond the Dateline. An explanation did not dawn on him until the extraordinary conditions of the 1982–3 El Niño when it became clear that in very strong events, prolonged westerlies can blow as far east as Easter Island and extend nearly all the way to the South American coast. Later, during the 1986 El Niño, when Finney accompanied the *Hokule'a* from Samoa to Hawaii via Tahiti, he discovered that even in mild events the abnormal westerlies reach at least as far east as Tahiti. Could the Polynesians have exploited the extraordinary conditions of El Niño's westerlies to make their incredible journeys?

In some cases, these anomalous winds may have taken islanders by surprise and blown canoes off course. Certainly, early European explorers recorded many examples of this. In 1826, for instance, the explorer F.W. Beechey chanced upon a group of forty castaways on a small island in the Tuamotu archipelago.[8] Two years beforehand, they had set out to sail the 300 miles westwards from their home on the island of Anaa to Tahiti, to attend the coronation ceremony of its new king. Coinciding with the 1824–5 El Niño, they were struck by a storm and strong westerlies and blown 600 miles in the opposite direction. When Beechey stumbled upon the group, they were marooned on a previously uninhabited island, still struggling to repair their canoes.

An even longer journey was recorded in 1951. Seven men travelling on business in the Marshall Islands were blown off course by a gale. As they strayed near Bikini atoll, radiation from the recent atomic bomb tests disabled their compass but they presumed the trades would carry them west to the Philippines. Instead, El Niño winds swept them south on a record 1,800 mile journey in just 106 days to the island of Epi in the New Hebrides.[9]

According to computer simulation drift experiments, such accidental voyaging is highly unlikely to account for the colonisation of the remotest islands.[10] In all reasonable probability, these trips could only have been achieved by those deliberately seeking new land and using their trained observations of wave patterns and bird flights to pinpoint their far-flung destinations. These skilled navigators probably also knew how to interpret the change in winds, sea temperatures and marine life that preceded an El Niño, and how best to take advantage of the persistent westerlies. Exploration to the west made perfect sense, because the prevailing trades allowed a relatively easy return journey with news of a discovery.

We can only guess why such risky adventures were undertaken. Perhaps it was due to a dearth of suitable land stemming from the Polynesian tradition of primogeniture, whereby all land is passed solely to the eldest son, or maybe it was just the simple lure of potential treasure. In many cases, exile was the inevitable consequence of the loss of wars and community conflict. But perhaps the most common trigger was the starvation and conflict stemming from the extreme droughts of El Niños and La Niñas. The result to some survivors was a loss of faith in the king's ability to make rain, or the island's ability to provide for the future. In fact, when Captain Cook and crew arrived in Hawaii, Cook's second-in-command, James King, found the locals

> imagined we came from some country where provisions had failed; and that our visit to them was merely for the purpose of filling our bellies. Indeed, the meagre appearance of some of our crew, the hearty appetites with which we sat down to their fresh provisions, and our great anxiety to purchase, and carry off, as much as we were able, led them, naturally enough, to such a conclusion.[11]

The irony is that the lack of rainfall that forced migration upon the islanders is likely to have been linked to the very conditions that enabled the Polynesian adventurers to reach way beyond what was normally possible in their long, upwind struggle to the east.

1520[12]

When Ferdinand Magellan first cast his eyes upon the Pacific, as he emerged from the South American straits that would later bear his name, all he could see ahead was an endless calm, blue horizon. He had absolutely no idea of the vastness of the ocean in front of him. The distorted concepts about the size of the world and of Far Eastern geography had left Magellan presuming he was no more than four or five days' sailing away from his destination of the Spice Islands. In reality, what lay ahead was one of the most harrowing tales of deprivation in the annals of nautical exploration. From November 1520, his crew endured three months and twenty days at sea without once finding a place to land before finally reaching Guam. According to Pigafetta, Magellan's chronicler, food stores were soon reduced to 'only old biscuit turned to powder, all full of worms and stinking of the urine which the rats had made on it, having eaten the said biscuit'.[13] By the time they crossed the equator, the crew were eating sawdust and the ox-hides that protected the rigging.

In contrast to the appalling conditions aboard, the ocean itself was wonderfully placid. Although Magellan's actual words were never recorded, it is known that his impressions inspired the naming of the ocean. His address to the crew has been surmised by one author as follows:

My comrades, we are sailing on an unknown ocean. No European ship has ever ploughed these gentle waters. On our charts this vast expanse is nameless. Do you not see how smooth as a lake is its surface, how mild are its breezes, how soft and even is its temperature. Comrades, I will give this great sea a name, and christen it. Henceforth let it be known as the Pacific.[14]

But were these conditions normal? Unfortunately, his energy having been sapped by fatigue and hunger, Pigafetta's diary of the journey contains little revealing detail. The scraps that are recorded about the winds note the ship spent most of the journey coasting under the steady breezes of the trades, so much so that 'for many days it was not necessary for the sailors to touch the helm or the sails'.[15] At first, the presence of the trades might suggest a 'normal' year, but even in El Niños, in weaker events particularly, the trade winds can still operate east of the Dateline. However, hidden between Pigafetta's lines there are some clues that they were sailing in an El Niño year.

A few days after entering the Pacific, near a point of land that is probably Cabo Tres Montes, on the tip of southern

Chile's Taitao Peninsula, Pigafetta noticed that the familiar sub-Antarctic species of fish were replaced by less familiar ones.

> In that Ocean Sea one sees a very amusing fish hunt. The [predator] fish are three sorts . . . and are called Dorado, Albicore and Bonito. Those fish follow the flying fish, called colondrini . . . When the above three kinds of fish find any of those flying fish, the latter immediately leap from the water and fly as long as their wings are wet – more than a crossbow's flight. While they are flying, the others [swim] behind them in the water, following the shadow of the flying fish. The latter have no sooner fallen into the water than the others immediately seize and eat them. This spectacle is a fine thing to watch.[16]

In normal years, the fish Pigafetta described would not be found in these waters. Each is a tropical species usually found either much closer to the equator or out in the deep ocean where upwelling does not cool the surface waters. In El Niño events, however, the warm water that moves eastwards across the Pacific spreads north and south once it hits the South American continent, and in its wake tropical fish extend their range down as far as Chile.

Magellan's vessel speed is potentially revealing too. For the first five weeks, having left the coastline in northern Chile where El Niño's influence on the trades would be expected to be weakest, the average speed was four knots. For the next section as they approached the equator, passing islands presumed to be Pukapuka and the Carolines, the speed slowed to 2.5 knots. From the equator to Guam, average speed ranged between five and six knots, and the final leg in March from Guam to Samar in the Philippines, took the amazingly fast time of eight days, at over eight knots.[17] There may be other reasons to account for this variation, but the timing of the improved speeds in the early months of 1521 does coincide with the most expected time of an El Niño's decay in the Pacific. And the exceptional speed of the final leg between Guam and Samar may suggest that by then the trades had been strengthened by a shift to La Niña conditions.

If Magellan did sail the Pacific during an El Niño, then undoubtedly the particular conditions of the ocean would have been unusual. Presumably, too, the trade winds were weakened, extending the hellish wait before the ship found land. Maybe if he had sailed in a 'normal' year, Pigafetta would not have made the slightly premature conclusion: 'I verily believe that nevermore will any man undertake to make such a voyage.'[18]

1532

It seems that from the very outset of Francisco Pizarro's expedition to conquer the Incas, he was beset with unusual weather conditions. After setting out from Panama in 1531, Pizarro reached southern Colombia in thirteen days, a trip that a few years earlier took him two years. By the time Pizarro had reached Ecuador's Guayaquil Bay, late in the year, heavy rains held him up and according to his personal secretary, Francisco de Jerez, he could not have advanced in the rains without serious detriment. By early 1532, Pizarro had reached Tumbes in northern Peru, where the river had increased in size and could not be forded. The hardest part of the journey lay ahead.

Between Tumbes and the Incan stronghold of Cajamarca lay 200 kilometres of coastal desert. With only two guaranteed sources of water in between, and over 160 soldiers and all their equipment to carry, this stage was potentially insurmountable. Yet, Jerez records, 'the governor [Pizarro] left the city of San Miguel in search of Atahualpa [the Incan king] on 24 September 1532. On the first day of the journey he and his men crossed the river on two rafts, the horses swimming'.[19] San Miguel lies on the Piura River, a stream normally dry between July and December, except in El Niños. Further south, Jerez also records, 'he [Pizarro] crossed a tract of dry, sandy country to reach a well-populated valley through which ran a large and swift river, which was in spate'.[20] Pizarro had inadvertently timed his desert crossing with the most favourable conditions he could have wished for. Not only was there an abundance of running water, it can also be assumed that the deserts were blooming with vegetation for the horses, and the desert communities encountered were relatively well stocked with food.

Pizarro was wonderfully lucky. His march happened at a time of civil war between Atahualpa and his brother for control of the Incan kingdom, thereby dividing the forces that were to oppose him. Just as Pizarro began his march, Atahualpa was camped at Cajamarca close to his route, saving Pizarro an exhausting trek down the length of Peru in pursuit of him. But whether El Niño's improved desert conditions were just another stroke of good fortune, we may never be sure. A few years later, one of Pizarro's party may have had a less fortuitous encounter with El Niño.

1539–40

Fray Marcos is one of the great mystery men of American history. For a year in his life, he became the most famous man in the whole of New Spain, as Mexico was known at the time. His adventures led directly to the greatest expedition of discovery in the history of America. Instantly he was hailed throughout Europe as a sixteenth-century Columbus. Then, just as quickly as he had found fame, he became a pariah, a laughing stock, the man dubbed the 'Lying Monk'. 'No man in history',[21] in the opinion of historian Hubert Howe Bancroft, 'has been so persistently slandered as Marcos.'

Marcos de Niza was a French friar of the Order of St Francis. In 1539, he was asked by Antonio Mendoza, the Viceroy of New Spain, to undertake a long journey from Mexico into the unknown lands to the north. Mendoza had received intriguing news of the existence of a region of seven cities, a number significant to many Europeans of the day. Legend had it that centuries before, seven bishops had escaped persecution in Spain and sailed west to establish the rich and beautiful Seven Cities of Antila. Although Columbus had failed to discover them, other mariners were purported to have uncovered concrete evidence of the mythical cities. Did this tantalising information also relate to them, and was their location finally to be revealed? Mendoza had to act fast, since he feared his rival Hernán Cortés, the conqueror of the Aztecs, might beat him to Antila. Without being able to obtain royal authority from the King of Spain for the colonisation of the region, Mendoza's solution was to clothe the expedition as a missionary quest. The man he chose to lead it, Fray Marcos, had already shown his evangelistic zeal on Pizarro's march. In April 1539, the barefooted friar set out from the Sonoran Coast in northern Mexico with a guide. When he returned three and a half months later, his story fanned the dreams of thousands across New Spain and Europe.

The Seven Cities, as Marcos described, were everything the legend had promised, and more. The particular town that Marcos claimed he had personally spied was Cibola, a great walled city 'bigger than the city of Mexico',[22] lined by long streets and plazas, and of tall stone houses with façades encrusted with rare gems. And this was reputed to be the most modest of the seven cities. As well as tales of fabulous

treasures, Marcos also described a journey through a wonderfully fertile desert. Along the way, the friendly inhabitants
had treated him with great kindness and generosity, supplying
him with more than enough corn, game and countless deli-
cacies. One valley was 'so well provided that it would suffice
to feed more than three hundred of horse: it is all irrigated and
is like a garden'.[23] Nothing more enticing could have been
wished for a potential new expedition.

In April 1540, a great military crusade set out under the
command of Francisco de Coronado. Marcos was invited
along as a guide. Coronado had not expected all his soldiers
and animals to live off the land, but he certainly hoped it
would provide some assistance. Yet, discrepancies soon arose
between Marcos's descriptions and what he was witnessing
for himself. For all Coronado could see, he might as well have
been travelling in another part of the world. At the same
points in the journey where Marcos had claimed that he had
been amply supplied with game, Coronado found the locals
unwilling or unable to provide anything. Food supplies became desperately low, and Marcos's word began to be
doubted. Coronado lost many men and animals through
starvation, and his remaining forces were badly weakened.
When the advance party reached Cibola, instead of returning
for reinforcements as planned, hunger forced a premature
raid. 'We were in such great need of food that I thought we
should all die of hunger if we continued to be without
provisions for another day,'[24] wrote Coronado.

As Coronado strode on to the plain before Cibola, his great
hopes for fortune and glory melted into disappointment. The
great silver city that he expected transpired to be nothing
more than a little Indian pueblo, still in existence today near
the borders of New Mexico and Arizona as the modest Zuñi
town of Hawikuh. 'There are cattle ranches in New
Spain,' moaned Coronado's chronicler, 'that make a better

appearance from a distance.'[25] Once inside the walls, after a battle of less than an hour, the soldiers gorged themselves on corns and beans from the storehouses. But the gold and precious jewels that were promised were nowhere to be seen. Marcos's reputation was in tatters, and he was banished on the first available convoy returning to New Spain. Later on, after exploring much of America's south-west and discovering the Grand Canyon, Coronado wrote of Marcos to the Viceroy: 'I can assure your lordship that he has not told the truth in a single thing that he has said, but everything is the reverse of what he said . . .'[26]

So why was everything the reverse? Was Marcos truly the 'Lying Monk' he so infamously became known as? It is impossible to believe that Cibola was transformed sufficiently within a year to excuse him. But what about the state of the country that lay between New Spain and Cibola, and the ease with which his party was able to travel in comparison to Coronado's? The seasons can't account for the difference, because both Marcos and Coronado's parties travelled at the same time of year. If there was an El Niño at the time of Marcos's journey, it could at least account for the discrepancies in the descriptions of the desert. But what is the evidence for this?

At the same time that Marcos was setting out on his journey, 30,000 Indians were reported to have starved in the Andean town of Cuzco after drought, a common occurrence there during El Niños. At the same time, Spanish counterparts in New Mexico were experiencing a winter of heavy snowfall and bitter cold, with the Rio Grande frozen solid for four months.

The news from Abyssinia was equally bleak in 1540. According to the official chronicle of the day, 'there was a great famine, the like of which had not been seen at the time of the kings of Samaria nor at the time of the destruction of the second temple'.[27] But by 1541, the indications are that conditions had switched to a La Niña phase. The Abyssinian chronicler was certainly much happier:

> The sky gave rain, and the earth produced all its fruits. The inhabitants who had nothing but the roots of trees to eat during the great famine, trampled on bread which was abundant as stones.[28]

The newly enthroned Emperor Galawdewos took the opportunity of renewed food supplies to begin the fightback against several years of Arab occupation, culminating two years later in the return of the country to Christian rule.

By the summer of 1541, little rain had fallen for over a year in both the north and south of China, with famine particularly severe throughout the north. Prince Altan, one of the many Mongol warlords who controlled the entire northern border, attempted to gain supplies by petitioning the royal court for special trading privileges. As the Mongols had not submitted tribute themselves for over forty years, the Minister for War refused. Prince Altan's response was to conduct a series of terrifying raids into Shanxi province, with much loss in life and property for the Chinese. For the next forty years, he remained the most feared and powerful of all the Mongols.[29]

At the same time in central America, the year 1541 was marked by high winds, frosts and torrential rains, a combination often associated with a La Niña. The rain fell so heavily on the slopes surrounding Santiago in Almolonga, the former Guatemalan capital, that a boulder-laden mud slide was washed down on to the town, killing over 100 Spaniards and several hundred Indian and African slaves. The town, which lasted just fourteen years in its original location, 'was left so damaged by the storm', according to the Bishop, 'that by necessity we moved it.'[30] By way of announcing their choice of a new location, the town's governors printed the New World's first newspaper.

Judging from the various sources of evidence from both the Old and the New Worlds, there appears to have been a strong and extended El Niño between 1539 and early 1541. This would certainly explain why Fray Marcos described the desert so favourably. And if by the middle of 1540 El Niño's influence on the weather of south-western America had changed from the previous year – a plausible scenario given El Niño's highly variable nature – it could help explain why Coronado found the land so contrastingly unhelpful.

While El Niño may have shown different faces to the two expeditions, it certainly cannot fully excuse Marcos. His descriptions of Cibola and its riches were clearly fabricated. In fact, most modern scholars who have examined his diary have concluded Marcos never got within days of Cibola. From their analysis of his journal, it seems highly likely that as Marcos neared the town, he received news that his guide, who had gone ahead to pave the way, had been slaughtered. Petrified, Marcos may have chosen to flee. Yet, the shame of being exposed as a coward and the guilt of not fulfilling the ultimate objective of his expedition may have forced him to invent parts of his story, drawing from what he had heard from villagers along the way.

His descriptions of the accommodating environment, on the other hand, may not have been so embellished. If the El Niño did bring rains in 1539, then Marcos naturally would have perceived the fertile landscape and favourable travelling conditions to be the norm, and the existence of the cities more plausible. It is interesting to speculate whether Coronado would have been so keen to dismiss Marcos as a 'Lying Monk' had be also enjoyed the same providence.

1578

By 1578, Francis Drake had already established himself as a respected admiral, but his circumnavigation of the globe in the *Golden Hind* that began that year cemented his reputation as the most renowned seaman of the Elizabethan age. Much of what is known about the expedition is recorded in *The World Encompassed*, the ship's journal by Francis Fletcher, the *Hind*'s preacher. One section, however, remains shrouded in mystery and controversy. Fletcher's description of the North American coastline throws up similar surprises to Marcos de Niza's unlikely account of conditions in south-west America, casting similar doubts over the integrity of Drake's chronicler. References to snow-covered hills in June, in particular, have since prompted accusations of deliberate falsehoods.

Having robbed and plundered Spanish settlements and vessels all along the Pacific coastline of South and Central America, Drake sailed north from Mexico in search of the mystical North-West Passage that would return him to the Atlantic and home. But even by the latitude of northern California, the crew was finding the cold unbearable. 'Though sea-men lack not good stomachs, yet it seemed a question to many amongst us, whether their hands should feed their mouths, or rather keepe themselves within their coverts from the pinching cold that did benumme them.'[31]

If Fletcher's diary is to be believed, the cold became a total preoccupation with the crew, and by 48 degrees north, roughly the Canadian–American border today, the search for the passage into the Atlantic was abandoned, and they turned south desperate for some warmer weather. 'To go further north, the extremity of the cold (which had now utterly discouraged our men) would not permit us; and the winds directly bent against us having once gotten under sayle againe, commandeered us to the Southward whether we would or no.'[32]

The situation further south, where they rested for two months during June and July of 1579 in a 'convenient and fit harborough', proved to be little improvement.

> It was in the height of Summer, and so neere the Sunne; yet were wee continually visited with like nipping colds as we had felt before . . . the generall squalidnesse and barrennesse of the countrie, hence come it, that in the middest of their Summer the snow hardly departed even from their very doores, but is never taken away from their hils at all; hence come those thicke mists and most stinking fogges . . .[33]

They also found the natives struggling with the cold. Historians are still arguing over the precise location of this arctic summer, but it is generally accepted to be within twenty miles of San Francisco's Golden Gate.

A number of possible explanations have been presented to account for Fletcher's reputed 'falsehoods'.[34] One suggestion is that the naval authorities back in England suppressed him from disclosing the 'exact truth' for strategic reasons. Another theory is that the descriptions of extreme weather conditions were designed to discourage Spanish exploration. Other critics simply doubt that Drake ever travelled further north than 43 degrees and that Fletcher invented the account of climatic conditions further north. In the early part of the nineteenth century, the American government even highlighted doubts over the *Hind*'s most northerly destination to strengthen their negotiating position over the delineation of the American-Canadian border. Few, it seems, have been prepared to take Fletcher's descriptions at face value.

In recent years, Fletcher's journal has been re-examined by a more open-minded audience. It has been pointed out that

Drake's journey occurred during the Little Ice Age when colder conditions could be expected. It is perhaps more pertinent that Drake's time in California coincided with the tail end of the strong 1578 El Niño and that by the summer of 1579 it may well have switched to a La Niña phase. The central Californian coast does not exhibit a consistent response to La Niña events, because it is situated between southern California, which does receive consistently lighter rainfall and higher temperatures, and the Oregon coast, which tends towards colder and wetter conditions. So it is not inconceivable then that if 1579 was a La Niña year, the arctic jet stream was pulled further south than normal across the Oregon and Californian coast, bringing heavy rain, storms and exceptionally cold conditions.

1587–9

It has often been said that the individual weather event with the greatest historical consequence was the storm of September 1588 that scuttled the Spanish Armada off the Irish coast, thereby saving England from invasion. Yet, from a body of evidence from around the globe, that North Sea storm was only one small part of a catalogue of extreme weather conditions around that period that had far-reaching historical consequences on four continents.

The first European to be born in America was a girl called Virginia Dare, who was born in July 1587 on Roanoke Island off the North Carolina coast. Sir Walter Raleigh had just begun the first permanent settlement of North America by establishing a party of British colonists to claim dominion over the American coast north of Florida, where the Spanish had already staked their claim. An all-male party had been left on the island two years previously, but fled back to Britain after harassment from the native Indians. The second attempt was altogether more serious. This time the 117-strong party consisted of men, women and children, provisioned with maps, seeds, agricultural tools, armour and cannons. However, examination of the stores upon arrival showed food would become precariously low, before the first crops could be harvested in the following spring. Leaving his daughter and grandchild on the island, John White, the settlement's governor, was apportioned the task of travelling back to Britain with Raleigh's boats and returning to the island immediately with additional stores.

The next few years were extremely cruel for White. Just as he arrived back in Britain, war with Spain was declared, and all major vessels and crews capable of providing him with a swift turnaround were being mobilised to stave off the threat of the Armada. Two small pinnaces were all that White could commandeer, but his desperate bid to cross the Atlantic was dashed in the Azores where he was overrun and pillaged by French pirates. As the months wore on, interest in the colony dwindled, and all White's efforts to generate concern for the plight of the colonists and his family were rebuffed. It was not until August 1590 that he finally made it back to Roanoke on a merchant vessel.

Finding the island deserted on his arrival was no particular surprise to White. When he originally left, plans were already being formed to move the settlement. If such a move was to be enacted, the instruction was 'in any wayes they should not faile to write or carve on the trees or posts of the dores the name of the place where they should be seated'.[35] And so it was that White found, engraved into the base of a tree, the word 'Croatan'. The Croatan tribe was based fifty miles to the south, and their chief was known to be friendly. Even with this information, White was out of luck. Following bad weather and the impatience of the sailors to continue their voyage, he was persuaded to postpone the journey to Croatan

Island until the following spring, and spend the winter in the Caribbean. But in sailing south, contrary winds blew the ship way off course, and White was forced back across the Atlantic, eventually returning to England. White's one opportunity had been lost. He later retired to Ireland, resigned never to see his family again. Despite further searches over the following years, no trace of the colonists was ever found.

What happened to the party of 117 has remained one of the enduring mysteries of early American history. One possible explanation for its disappearance was uncovered in the late 1990s when tree-rings from south-eastern United States were examined to establish a climatic chronology for the region.[36] The evidence shows that the years 1587–9 represented the most extreme drought over the last 800 years for the entire east coast. Before he left the colony in 1587, White witnessed a small skirmish between the settlers and native Croatans over crops, possibly a hint of the onset of drought. Could El Niño have played a role in the disappearance of the settlers? While abnormal rainfall on America's central-east coast is often associated with El Niño and La Niña events, it is not in itself definitive proof of El Niño activity. The North Atlantic Oscillation and Atlantic sea-surface temperature fluctuations can account for much of the variability in the region's weather conditions. Indeed, Quinn's time-series does not recognise an El Niño until 1589. However, there is no shortage of evidence of unusual weather conditions around the world between 1587 and 1589 to suggest one.

According to reports from European missionaries and travellers, 1587 saw drought in Mexico, Vietnam and in Brazil where several thousands of indigenous tribals were forced to migrate to the coast. The next two years brought terrible weather across Europe, leading to famines and food shortages in England and France. In Ireland, conditions were so bad that one chronicler wrote, 'one did eate another for hunger',[37] while another in Iceland described the misery as 'this most terrible visitor'.[38] All over the Atlantic, an exceptional number of hurricanes and storms destroyed hundreds of Spanish ships. In two separate regions, as yet untouched by European expansion, the extreme weather had a profound influence on history.

The puzzling events on America's East Coast also coincided with a tremendous upheaval on the African continent, where millions of people were affected. In East Africa, on an island in the Albert Nile just north of Lake Albert, a tree still grows whose origins are said to date from those years, symbolising the tumultuous events that engulfed much of the continent during the period. According to the story still told by the Luo people of western Kenya and northern Uganda, it was from here their tribe fled famine, splitting into two in southern Sudan. Tribal law decreed that neither side was ever allowed to cross the Nile again, and 'those who disregarded it should die'.[39] To mark the acrimonious separation, an axe was driven into the dry river-bed of the Nile, from where, it is said, the axe handle grew into the tree that still grows today. The location is known as Wat Latong, meaning the 'Port of the Axe'.

In this epoch-making event in east African history, one group moved west to establish dominance over Sudanic peoples. The other moved south along the eastern shores of Lake Albert, establishing hegemony over the Bunyoro State, and founding the vast Bito Dynasty that stretched north to Ethiopia, to the Congo in the west, and the shores of Lake Tanganyika in the south. The Luo have been very successful in spreading their language and culture, yet theirs was just one small movement in a seismic reshuffle of power and peoples.

To African historians, the time is known as the *Nyarubanga*, from the Luo term meaning 'daughter of God', which in the context of famine denotes 'one sent by God as punishment'. Through the tracing of royal genealogies, and the recording of one of the lowest Nile flows at Cairo, the *Nyarubanga* has been dated to between 1587 and 1589. The same year, a Dominican priest travelling along the east coast witnessed the aftermath of an appalling locust plague.

> There was so great a scarcity of provisions that the kaffirs came to sell themselves as slaves merely to obtain food, and exchanged their children for a alqueire of millet, and those who could not avail themselves of this remedy perished of hunger, so that at this time a great number of the inhabitants of these lands died.[40]

The drought moved one contemporary historian to write, 'The stark spectre of starvation leaps from the traditions of this period as never before or since.'[41] Drought extended throughout south and eastern Africa, and more patchily north and west.

The appearance of drought unleashed political tensions and fuelled inter-ethnic rivalries. Anarchy, starvation and marauding armies are recurrent themes in the oral histories relating to the period. Cannibalism, too, was said to be rampant, although very uncharacteristic for the people of the region. Many dynasties and tribes can relate their histories back to this period, often beginning with a hero-founder, leading his people away from famine to settle in sparsely populated areas.

Out of the turmoil came an unprecedented redrawing of the political map of central, eastern and southern Africa. Major migrations, invasions and revolts typified the response, each triggering a domino reaction that reverberated through much of the continent. Throughout the East African lakes region there was a general southerly movement of states and chiefdoms. In addition to the Luo migration, the Paranilotes –

ancestors of the Masai, Turkana and Samburu peoples –
moved into the Baar region of southern Sudan, pushing the
resident Madi further south. The starving Bunyoro fled their
home in what today is western Uganda, invading the Rwanda
Dynasty and causing its collapse. As an indication of the state
of the Victoria Nile at this time, the Bunyoro king who
crossed it earned the nickname *Kyambukyankanko* – 'the
one who crosses a river with shoes on'. The Babinza fled east
across a dry Smith Sound in the southern arm of Lake
Victoria, to settle in Sukuma.

Further south, drought and famine prompted more migrations. The infamous cannibals of the Jaga and Zimba tribes
were on the move 'eating every living thing'.[41] The Zimba
emerged out of the Maravi empire in the Malawi/Mozambique
area, ravaging the Zambezi Valley and Portuguese trading
towns of the east coast. In fact, it was the Zimba's rape of
Mombasa in 1589 that enabled the Portuguese to occupy this
outpost of Muslim resistance. Meanwhile, the Jaga were
rampaging through the Angola and Congo areas. Drought
also saw the break-up of the Sotho-Tswana Dynasty in the
Limpopo Valley.

In West Africa, the drought-weakened Songhai empire
suffered civil war and the collapse of its government. Its
helpless state left it open to defeat in 1591 by a greatly
out-numbered Moroccan army that had just struggled across
the Sahara.

It is estimated that the upheavals that began during the
Nyarubanga and continued during the unusually dry decades
that followed, culminating in the Great Famine of 1620–1,
saw the death of half the entire population of East and central
Africa.

The far-reaching consequences of climatic abnormalities
were also felt much further afield. The same period saw major
demographic change in China. In 1587 major floods in the

Yangtze River Delta, and heavy rains at harvest time, devastated crops. Further north drought reigned. By 1588 a major
drought stretched all the way from Shandong and Shanxi in
the north at least to Zhejiang in central China, and continued
into the following year.

The natural disasters marked a period of civil unrest with a
series of demonstrations against 'gluttonous officials and
corrupt clerks'. It has been noted that the people 'all began
to stir at once, acting together without [prior] agreement, and
in the people's hearts there was a change in manners and
morals'.[42]

In the midst of a series of natural calamities, one of the two
most widespread and lethal epidemics of China's history
erupted. Significantly, the second outbreak occurred during
the other great climatic calamity of that era in the El Niño of
1640–1, which contributed to the fall of the Ming Dynasty, as
we have seen. Fanning out like ink on blotting paper, both
outbreaks moved along the arms of the Yangtze and Yellow
Rivers, as well as the Grand Canal, embracing most of
China's densely populated areas. The epidemics also corresponded to areas hit by climate-induced famine. Although
climatic conditions are often associated with an increase in the
spread and virulence of a cocktail of diseases, including
typhoid and dysentery, the main agent of these Chinese
epidemics has remained a mystery. Plague is a distinct possibility and, as can be seen in the next chapter, there is a
suggested historical correlation between plague outbreaks
and El Niño events. Another suggestion springs from the
term 'big-head fever' that was used at the time, possibly
describing hydrocephalic meningitis. At its peak, the epidemic
is recorded as claiming 80–90 per cent mortality which, if
accurate, features a virulence virtually unknown in epidemiological history and suggests more than one causal factor. It is
no wonder the disease was also referred to as 'peep-sickness',

because of the belief, 'You only had to peep and you would catch the disease and die.' Between the two outbreaks, some estimate that between 20 and 40 per cent of China's total population died. Others put the figures even higher. The problems of over-population experienced in the late Ming Dynasty were not repeated for another 150 years.

1652[43]

From Magellan's time, any vessel trading with the Spice Islands travelled via the southern tip of Africa. Many of those vessels stopped in Cape Town's sheltered bay to trade tobacco, metals and beads for the cattle of the local Khoi people. After glowing reports of the favourable climate for growing crops, the Dutch East India Company decided to settle a permanent station to allow the Company's trading boats to add fresh fruit and vegetables to the supplies.

Arriving in mid-1652, just in time for the annual summer rains, the colonists must have expected the right conditions for their first crop. But 1652 was an El Niño; typically in these years, rainfall is much higher than average. In 1652, the rains were heavy and prolonged, flooding the valley below Table Mountain and destroying the new gardens. To add insult to injury, the valleys were now swarming with hartebeest and eland, attracted by the flush of new vegetation, and the colonists' primitive firelocks proved incapable of bringing any of them down. Starvation and dysentery brought on by the wet conditions followed, leaving over half the original party too weak to work. Jan van Riebeeck, the colony's first governor, wrote:

> As a result of this inclement weather almost all our crops have been destroyed, and at the same time we are entirely prevented from proceeding with the necessary work. The men also are

getting sicker and weaker every day through the discomfort
they are suffering. Nothing can be done about it but to wait
for better and more favourable weather; for this patience will
have to be exercised in the hope that the Almighty will provide
it.[44]

Salvation appeared a few months later, not for the last time in
South Africa's history, from the unlikely outpost of Robben
Island. Here, an abundant supply of penguins, seals and
nesting birds proved just enough to sustain the colony until
better conditions arrived the next year.

1769–70[45]

In 1769, a fledgling community on the west coast of America
had a very similar beginning. The mission of San Diego de
Alcalá, the first settlement of upper California, was established by a small group of Franciscan missionaries who had
marched north from the Gulf of California. After surveying
many prospective locations, a site on the north side of San
Diego Bay, familiar to residents today as the Old Town, was
selected for its agricultural prospects. Unfortunately for the
colonists, settlement coincided with an El Niño, the same
event that saw 10 million die from starvation in Bengal.
Heavy flooding occurred on several rivers in California that
winter, and in San Diego all crops planted along the broad
river plain were ruined. Starvation claimed the lives of 50 of
the 126 original settlers.

Efforts to convert the indigenous population were proving
equally unrewarding. Suffering from hunger, the emaciated
colonists made a pathetic spectacle. Despite their best efforts
to contact the indigenous tribes, the visitors' bloated faces and
swollen limbs presented Christianity in a most unappealing
light. If this was all the new God offered, the locals preferred

to stick to their own. In the first year, there was not a single conversion, and all talk centred on whether to abandon the new colony.

Just in the nick of time, a relief vessel arrived in March 1770, laden with supplies. With improved weather later that year, vines and fruit trees grew abundantly, and local Indians began slowly to be persuaded to the new faith. Eight more missions between San Diego and San Francisco followed over the decade and another ten before the turn of the century. The Europeanisation of America's west coast had begun.

Meanwhile, in Tahiti, where Lieutenant James Cook had landed the *Endeavour* in April 1769 to study the once-a-century transit of Venus across the sun, unusual conditions were also prevailing. 'The meeting with Westerly winds within the general limits of the easterly trade is a little extraordinary,'[46] he wrote in his journal. Occasional westerly winds are observed in Tahiti around the end of the year, but if occurring in other months they are usually associated with El Niño years.

Although the unusual meteorological conditions were of no great importance to Cook during his time in Tahiti, they certainly became so later on in the *Endeavour*'s journey. From Cook's journals it can be assumed that much rain had recently fallen around Botany Bay by the time he landed in April 1770. The streams were flowing swiftly, and everywhere grew 'vast quantities of grass', one meadow as fine 'as ever was seen'.[47] One of the streams was so swift-flowing, according to Joseph Banks, the ship's botanist, that it reminded him of water-driven flour mills back in England. Such a favourable impression was gained, according to the eminent historian Geoffrey Blainey, that it was 'on that misconception the British were to plant their first settlement.'[48] Yet when the first colonists arrived eighteen years later, the fields of grass had disappeared and the streams had become trickles. Blainey interpreted the lush conditions as being a product of seasonal rains, but it is quite possible, considering the flooding experienced in Timbuktu and Mexico, that 1770–1 had switched to a La Niña event. In fact, when the *Endeavour* sailed on to Java from Australia later that year, an unusually early monsoon took the locals by surprise. 'The quantities of bedding that I everywhere saw hung up to dry made a very uncommon sight,' wrote Joseph Banks. 'The people here told us that it did not commonly shift so suddenly, and were loth to believe that the westerly winds were really set in for several days after.'[49]

The consequences of the extra rain around Botany Bay were by no means trivial. It was an important expectation of the British government that tropical produce from the new colony would soon be able to substitute that imported from Dutch and French colonies. Were it not for the unusual state of the land when Cook arrived, as Blainey has concluded, the British 'probably would have rejected the continent'.[50]

1791–2[51]

In April 1793, the French explorer D'Entrecasteaux landed at Balade, on the north coast of New Caledonia, the same spot where his English counterpart James Cook had set foot seventeen years before. Throughout his Pacific travels, D'Entrecasteaux had found Cook's journals a valuable tool for navigation, as well as an accurate and objective guide to the peoples and food sources of the various islands. But Cook's descriptions of the natives of New Caledonia caused D'Entrecasteaux to wonder if Cook really was referring to the same island.

Cook, not one for sentimentality or falsehoods in his diaries, couldn't praise the islanders enough: 'They are a strong, robust, active, well-made people, courteous and friendly and not in the least addicted to pilfering . . . In their disposition, they are like the natives of the Friendly Isles; but in affability and honesty they excel them.'[52] Immediately, a rapport was built up whereby the natives wandered freely aboard the ship and the Europeans were free to explore around the island at their leisure. 'They had little else than good nature to bestow,' Cook concluded.

The reception extended to the French was in stark contrast. Instead of being offered coconuts and yams, they were shown clubs and slings. Far from 'behaving with the strictest propriety and honesty'[53] as they had read, the thieving became so bad that the French were forced to draw a line in the sand, with their stocks on one side, and warn of the direst consequences should the mark be overstepped. Judging from the abundance of recent wounds and scars on their bodies, as well as the sight of columns of men with weapons and flaming torches marching into the hills at night, the local tribes seemed to be in conflict with one another.

As for the women, Cook had found them flirtatious but chaste, concluding with an observation highly unusual in the annals of early Pacific exploration: 'I never heard that one of our people obtained the least favour from any of them.'[54] To the French, the women were immodest and willing to sell themselves at any price.

One custom, most alarming and disconcerting to the French, had not been observed by the English at all. The first to notice it was Jean-Julian La Billardière, D'Entrecasteaux's botanist, as he was about to gnaw on a bone that had been passed to him in an uncharacteristically friendly gesture by a native. The bone he recognised, from his days as a medical student, as being that of a human child. Initially the sailors did not believe him, 'for they could not be persuaded that these people, of whom Captain Cook . . . had drawn so flattering a picture, were degraded by such a horrible vice'.[55]

However, far from the islanders being ashamed about the origins of the bones, 'they made no scruple to avow that the flesh which had covered them, had served as a meal to some islander; and they gave us to understand that they considered it a choice dish'. The locals' insistence that it was only the bones of enemies that were sought after did little to reassure the sailors.

Various propositions have been offered by historians as to how, only seventeen years apart, two such diametrically opposed impressions were reached. One explanation is that the two parties had encountered two different tribes, after migration had taken place in the interim. However, there is no evidence for this. Another explanation has centred on the cultural differences between the two sailing parties and their captains. Cook's visit was much shorter than D'Entrecasteaux's, possibly leading to a more superficial impression. In addition, Cook's nature and background may well have predisposed him to a more naïve and glowing interpretation of native virtues – an argument not substantiated by his descriptions of other islands and their tribes, accounts which have been adjudged elsewhere as objective and unromanticised. On the other hand, Cook's warm and diplomatic approach, including the respect he showed to the tribal chief, was sure to elicit a much more favourable response from the locals than the wary and aloof approach of the French. Still, the interaction between the locals and the two parties of whites can only ever be a partial explanation. Clearly, the situation on New Caledonia of 1793 was a very different one from that of Cook's visit. But it also coincided with the end of a very severe and extended El Niño, probably the strongest event of the eighteenth century, making drought a distinct possibility.

From the French accounts, that is exactly what seems to have happened. 'With one-hand they would reach out for worm-eaten biscuits,' recorded botanist Anders Sparrman in his memoirs about the island, 'while the other they pointed to their pinched-in stomachs, which they dragged still further in with their muscles to indicate hunger and arouse pity.'[56] In the mountains, where the inhabitants did not have direct access to the fruits of the sea, there were scenes of even greater misery. Some people were observed filling their mouths with steatite, an impure form of talc, which served, according to La Billiardière's medical knowledge, 'to deaden the sense of hunger by filling their stomach, thus supporting the viscera attached to the diaphragm; and although this substance does not afford any nutritious juice, it is yet very useful to these people'.[57] But if El Niño's imprint remained hazy on New Caledonia that year, elsewhere it is a lot more obvious.

On the other side of the Pacific, El Niño conditions in Hawaii revealed a hidden side to the highly accomplished British navigator George Vancouver. A former midshipman with Captain Cook and witness to Cook's death, Vancouver was entrusted with mapping the north-west Pacific coastline of America. In October 1792, he docked at the busy trading port of Nootka, situated on the island that would later bear Vancouver's name. Just as he was about to depart, he was alerted to the plight of two teenage Hawaiian girls, Raheina and Tymarow. A year beforehand, the girls had been seduced aboard a passing British ship, confined below deck until after setting sail and, according to rumour, were now in Nootka to be exchanged for furs. Vancouver, an austere and puritanical man, assumed custody of the girls by agreeing to return them to their native island, where he had already intended to winter to negotiate the cession of Hawaii to the British.

But if after several further months of surveying the

Californian coast there was great excitement on the part of
the girls as they neared their homeland, this soon turned to
disappointment. As the ship docked at the neighbouring
island of Kauai, they learnt that the inhabitants of their
own island of Niihau 'had almost entirely abandoned it, in
consequence of the excessive drought that had prevailed
during the last summer'.[58] His duty not yet discharged,
Vancouver faced a quandary over the girls' future. But reading between the lines of his journal, it was more than a sense
of obligation that stirred him.

During the months at sea together, the younger Raheina, in
particular, seems to have moved him in ways that his man-
nered prose found hard to contain.

> The elegance of Raheina's figure, the regularity and softness of
> her features, and the delicacy which she naturally possessed,
> gave her a superiority in point of personal accomplishments
> over the generality of her sex . . . a degree of personal delicacy
> that was conspicuous on many occasions . . . her sensibility
> and turn of mind, her sweetness of temper and complacency of
> manners, were beyond any thing that could have been ex-
> pected from her birth . . .[59]

Unable to return his captivating passengers to their now
uninhabited island, Vancouver went to extraordinary pains
to secure their future on Kauai, a larger island that had
escaped the worst of the drought. He begged the local chiefs
not to punish the girls for the 'heinous' crime of 'living in the
company of men' and received assurances that the many gifts
they had been given would not be taken away. He even
persuaded one chief to part with a large piece of land in a
fertile valley of the Waimea district. When Vancouver finally
bade his last affectionate farewell, the girls were ensconced on
their private estates reflecting on 'the various turns of fortune

that had conspired to place them in such comfortable circumstances'.[60]

The 1790–3 event was no ordinary El Niño, and disasters as well as profound political and social changes were triggered in many places around the globe.

Over one million died in the Indian province of Baroda through drought that spread to other parts of the country over the following two years. 'The season continued alarmingly dry – the crops of potatoes failed . . . and numbers of the cattle have died',[61] a report by the East India Company lamented over its interests. As for the people, the hunger was so widespread from the failed monsoon, particularly in the south, that suicide was prevalent and 'some killed their children and lived on their flesh.' The years 1791–2 became known as the Skull Famine, because the dead were too numerous to bury.

Meanwhile in Australia, a drought that began in 1790 dashed all efforts at establishing crops on the banks of Sydney Cove just two years after the first convicts arrived. The Tank Stream, the settlement's most immediate freshwater source which still runs underneath Sydney's city centre today, dried up completely, something that has never happened since. Arthur Phillip, the colony's first Governor-General, wrote as relief supplies slowly made their way from Cape Town, 'So little rain has fallen that most of the runs of water in the different parts of the harbour have been dried up for several months.'[62] Without foresight into the profound impact that future El Niños would have on the colony's history, he concluded, 'I do not think it probable that so dry a season often occurs.'

As in Australia, the El Niño seems to have impacted on northeast Brazil by late 1790. Here, over half the cattle herds were wiped out, taking decades to recover. The vital export trade in salt-beef from the inland provinces to coastal settlements was lost to suppliers from southern Brazil, and was never regained.

In Mexico, the towns of San Pedro Bartolomé and San

Pedro Pareo were embroiled in a legal battle over ownership of new land that had emerged from the drought's effect on lake-levels. There were droughts as well in parts of Chile and Russia. Iceland had most of its coastline surrounded by sea-ice until after the summer of 1792.

In Arizona, a series of floods permanently changed the course of several rivers after destroying towns, bridges and orchards, and even carrying off one 3,500-pound safe from a local brewery. More than a hundred settlers of the Gila River's lower reaches were feared drowned, but 'nearly all of them had escaped with their families by constructing rafts from the timbers of their houses and rowing to the hills'. Likewise in Peru, the rain never seemed to stop. 'No one now living recollects a rainy season anterior to that of 1791,' wrote one resident in Peru's north, 'when the river Piura rose so high that it flooded the principal plaza of the city.'[63]

When drought in South Africa forced the Xhosa to begin settling west of the Fish River in search of pasture for their cattle, Boer commandos attacked, instigating the Second Frontier War. The Xhosa retaliated fiercely and drove the Boers out of the *zuurveld*, a key fifty-mile strip of valuable coastal land that was later to become known as Albany.

1812

At the end of 1811, the *zuurveld* still remained in Xhosa hands. Against the background of drought, which once again exposed Boer insecurities over the productivity of the land on their frontier, the British government mounted a serious effort to push the Xhosa back east of the Fish River, culminating in the Fourth Frontier War. 'We chose the season of corn being on the ground in order . . . that we might the more severely punish them for their many crimes by destroying it,'[64] said a British officer.

Alexander Humboldt, who had just travelled through South America on one of the world's greatest explorations of natural history, also noted extreme weather conditions. 'A great drought prevailed at this period in the province of Venezuela. Not a single drop of rain had fallen in Caracas, or in the country ninety leagues round, during the five months which preceded the destruction of the capital [by earthquake].'[65] The mention of both natural disasters together was not coincidental. In parts of Latin America, a belief persisted that there was a correlation between droughts and earthquakes, because droughts were associated with less lightning, and the rarity of electric explosions led to 'an accumulation of electricity in the interior of the Earth'. The events of early 1812 in Venezuela, in the eyes of the locals, only served to compound this belief. Humboldt, however, was not convinced that the two events were anything more than a coincidence.

Great droughts were raging elsewhere too. In central and southern India, a famine was so severe that 'even people possessed of ample means, and otherwise in favourable circumstances, died from want of that grain which their rulers could not purchase'. In some parts 'mothers struggled with husbands to rescue innocent and helpless children from being killed by their fathers to appease uncontrollable hunger'.[66] In adjacent Burma, the year is known as *maha-thayawgyi*, the 'Great Famine', 'still remembered in many a northern village as a time of terror and starvation'.

By 1812, the fledgling colony of Australia was still no greater in size than the land on which Sydney stands today. After little more than two decades of growth, the settlement clung to the land surrounding Port Jackson Bay, boxed in by a line of hills to the north, south and west. None of the colonists knew what lay beyond. To the north lay China, many of the convicts presumed, and to the south existed another colony of whites 'in which they were assured of finding all the comforts of life, without the necessity of labouring for them'.[67]

By February 1812, several months of severe drought had exposed the limited stock-carrying capacity within the settlement's boundaries. When showers finally arrived, the much-awaited new grass that emerged was immediately ravaged by millions of Army-worm caterpillars. To Gregory Blaxland, one of the colony's most prominent and respectable graziers, the need for new land was desperate. His fiery temper had tarnished his relationship with the colony's Governor, who refused to grant Blaxland further grazing rights. The only new pasture lay beyond the mountains.

Blaxland, along with several others, set out in 1813 to cross the Blue Mountains, keeping to the ridges and avoiding the deep valleys wherever possible. On the other side they found the vast western plains, with enough grass to be 'equal to every demand which this country may have for an extension of tillage and pasture lands for a century to come'.[68] In an instant, the visions of the colony were transformed from a small coastal settlement to a vast and fertile continent.

The most infamous weather of 1812, however, was recorded in Russia. 'It is the winter that has been our undoing. We are the victims of climate,'[69] Napoleon wrote famously as he fled from Russia, leaving behind the death-throes of his *Grande Armée*. So began, in popular legend at least, the belief that his defeat was inflicted by 'General Winter', a reference to the exceptional cold that befell his troops in the last weeks of 1812. At the end of the campaign, less than one-tenth of the 675,000 men that Napoleon had led into Russia made it back across the border. It was a defeat of such colossal proportions that Napoleon's reputation never fully recovered. But how much was the El Niño to blame?

Even if the extreme winter that began in Russia in late 1812 was among the long catalogue of El Niño's impacts that year,

Napoleon had little justification in using it as an excuse. The freezing temperatures only set in beyond mid-November. By then, Napoleon was retreating from Moscow, his troops already starving and demoralised. The extraordinary weather of 1812 certainly killed, at the very least, tens of thousands, but as a cause of defeat it was no more than a scapegoat for his strategic blunders. Hitler on the other hand, who marched on Russia on almost the same day 129 years later, could, as we shall see a little later, blame El Niño for his failure with a lot more justification.

1844–6

In February 1846, a special meeting of the South Australian Royal Geographic Society was convened in honour of the safe return of Charles Sturt from a two-year expedition into the Australian wilderness. When Sturt finally arose to address the packed hall, he broke down in tears. Memories of the death of one of his party and the unspeakable hardships they endured were too much for him. In his eyes, the expedition had been a total failure, stopping 150 miles short of his main objective to reach the geographic heart of the country. Yet, importantly, he had managed to shatter the popular myth that the entire centre of Australia was covered by an enormous inland sea. Sturt's sole mistake was to inadvertently plan his trip almost perfectly to coincide with the great El Niño of 1844–6, the same one that arguably triggered the Irish potato famine.

Along with a party of fifteen others, plus horses, cattle, sheep and a boat with which to cross the inland sea, Sturt set out from Adelaide in August 1844, just as the bite of the 1844–6 El Niño drought was beginning to take hold. After striking camp in January near what is now the border of western New South Wales, reconnaissance trips failed to show water ahead. For six months of intense heat, with his small party dug in, Sturt described how 'the lead dropped out of our pencils, our signal rockets were entirely spoiled; our hair, as well as the wool on the sheep, ceased to grow, and our nails became as brittle as glass'.[70] When he later came across the watercourse that he named Cooper's Creek, a stream that in years of good rains can stretch as much as thirty miles wide in flood, he said: 'I would gladly have had this creek down as a river, but as it had no current I did not feel myself justified in so doing.'[71]

With the cold winter nights upon them, and the time of expected rains approaching, the party pushed on once again. Confident that they would soon be upon the inland sea, Sturt even ordered their boat to be painted. Yet, the further inland they pushed, the more infrequent the waterholes became and the more desperate the need for rain. 'I could not but hope,' Sturt wrote on seeing the rare sight of a potential raincloud, 'that the great author of it would have blessed the land, desert as it was, with moisture at last, but I listened in vain for the pattering of rain, no drops, whether heavy or light, fell on my tent.'[72]

By October 1845, Sturt stood facing a seemingly infinite series of parallel sandy ridges, rippling off past the horizon. Dangerously weakened by thirst and scurvy, he ordered his men to turn back. Now began a life-threatening race to reach each drying waterhole before it disappeared completely. 'The drought had now continued so long, and the heat had been so severe, that I apprehended we might be obliged to remain another summer in these fearful solitudes.' It was a country, so he concluded, 'such as I firmly believe has no parallel on Earth's surface.'[73]

To the many pioneering colonists with dreams of the fabulous wealth to be made in Australia's fertile centre, Sturt's descriptions of the heat and deprivation dashed all hopes. Upon hearing them, it was said, his wife's hair turned grey overnight. Soon afterwards, Sturt retired back to England, unable to judge his expedition as anything but a personal failure. He could hardly have imagined that a few decades on, the same 'fearful solitudes' would support flocks of sheep.

In parts of southern Africa, the El Niño was creating another kind of personal battle. Like Sturt, David Livingstone chose the same inopportune moment to begin his own crusade into darkest Africa. Having previously spent four years as an apprentice missionary in the Cape Colony, Livingstone began his own mission in 1845 among the Bakwena people at Chonuane, near today's border of Botswana and the Transvaal. But just as he arrived, so did a long drought. 'The clouds went round and round us until the people were saying "these clouds make sport of us",' he recorded in a letter. 'Our house was supposed the cause why no rain came down, and we were requested to allow them to sprinkle it with medicine. To this we had no objections, provided the stuff did not smell badly, yet no rain came.'[74]

In times of severe drought, the tribe's response was to call in a rainmaker – in Livingstone's words, a man 'most insignificant in appearance'. A clash of wills erupted immediately. Livingstone felt the rainmaker was fooling the people with his preparations of burned bats and the bowels of old cows, warning him to give up 'the folly and wickedness of his ways'. The rainmaker, in return, talked of the link between Livingstone's arrival and the onset of drought as no mere coincidence. 'White men know how to make guns,' the chief pleaded with him, 'and black men know how to make rain, and the latter ought not to be interfered with.'[75]

With no signs of the drought abating, Livingstone chose to move in 1846 to nearby Kolobeng, reputed to have a more reliable climate. But for another three years drought continued, even after the El Niño had decayed, with Livingstone complaining all the while of 'Satan's powers over the course of winds and clouds'.[76] Yet he was never able to remove the doubt in the minds of the Bakwena that the cause of the drought was himself. In 1850, he moved on in search of a less drought-prone region, having achieved in his own words 'very limited' results.

As an historical figure, Livingstone's ground-breaking reputation rests largely on his geographical discoveries, as well as his efforts to abolish the slave trade. Yet as a missionary, he never really recovered from his inauspicious start. Throughout his entire stay on the continent, the number of natives converted to Christianity totalled one, and even he lapsed back to heathenism.

1888

During the southern hemisphere summer of 1888–9, while a third of the Ethiopian population was dying from the consequences of drought and rinderpest, the weather on Samoa had been unusually fine, a 'circumstance that had been commented on as providential',[77] remembered Robert Louis Stevenson. Unfortunately the various foreign visitors to the island were not so attuned to the signs.

For some years, Germany had been manoeuvring against Britain and America for a monopoly of the island's trading interests. By late 1888, relations between the opposing sides had deteriorated so badly that all three countries had dispatched forces to the islands, and the threat of military action, on Samoa at least, was hanging in the air. By 15 March 1889, in an extraordinary case of brinkmanship, Samoa's tiny harbour of Apia was crammed with seven warships and numerous associated merchant craft.

The cyclone season on Samoa falls between November and April. As a rule of thumb, the islanders can generally expect to receive a cyclone every seven to ten years, a frequency that roughly corresponds to that of moderate to strong El Niño years. Although cyclones can occur at other times, El Niño's movement of warm water eastwards across the Pacific shifts the zone of likeliest cyclone formation to encompass Samoa.

When gale-force winds hit the harbour on 15 March and barometers dropped to 29 inches, the American admiral issued signals for all warships to head to sea. Not on speaking terms with his opposite number, the German commander ignored the signal. With no captain prepared to leave harbour without a guarantee the enemy was following, all stayed. During the night, the gale turned into a cyclone, and massive waves roared through into the funnel-shaped harbour. 'Seas that might have awakened surprise and terror in the midst of the Atlantic,' wrote Stevenson, 'ranged bodily and almost without diminution into the belly of that flask-shaped harbour, and the war-ships were alternatively buried from view in the trough, or seen standing on end against the breast of the billows.'[78]

The first casualty was the German gunboat *Eber*, sucked underneath the overhang of a reef. Anchors dragging, other ships drifted helplessly, with several German and American craft being crushed against each other. Only the British cruiser *Calliope*, with its superior power, managed to avoid hitting its drifting brethren and make it out to sea. All German and American boats were lost, and despite the brave efforts of native rescue parties, well over a hundred seamen died.

'Within the duration of a single day,' wrote Stevenson, who was greeted by the wrecks as he sailed into Apia harbour later that year, 'the sword-arm of each of the two angry powers was broken; their formidable ships reduced to junk; their disciplined hundreds to a horde of castaways.'[79]

Afterwards, the popular interpretation of the hurricane was that of an act of divine intervention, because any prospect of war on Samoa had literally been blown away. Later that year, a treaty was signed by the three countries dividing control of the island between them. The United States navy, according to a newspaper report of the time, was left with 'almost no . . . war vessels worthy of the name in the Pacific'.[80] It was just the ammunition necessary to combat opposition in Congress to naval modernisation, and the incident became widely credited as an important spur in the subsequent programme of American naval reconstruction.

'The loss of three American vessels,' complained a *New York Times* editorial, 'with so many of their officers and crew, is a national damage greater than there is any reason to expect that all the trade we can do with the islands in the course of the next half century can at all compensate.'[81]

1891–2

In the annals of Australian exploration, David Lindsay does not attract more than an obscure footnote. As a surveyor and expert camel handler, Lindsay had the right combination of technical and practical skills to be chosen in 1891 to lead a scientific expedition, considered to be the strongest and best equipped of its day. The goal was to explore parts of southern and western Australia, regarded as the last great blank space on the colony's map.

In May 1891, just as drought was beginning to take a grip on much of the continent, Lindsay marched west from Warrina in South Australia on an expedition that would go on to traverse over 6,400 kilometres. But even before the expedition passed the point previous explorers had reached, known waterholes were found choked with sand. With up to three weeks' travel between water-breaks, the camels were on the point of dropping. The drought-ravaged land was taking its toll on mind as well as body, and ill-feeling among the party was brewing towards Lindsay. By January, all the scientists except for second-in-command Lawrence Wells had resigned. Lindsay, in response, made a 500 kilometre dash to the coast to obtain further instructions from his financiers, only to discover he had been relieved of his post.

Wells was to be a lot more fortunate. The weeks following the take-over of command coincided with the decay of the El Niño in the Pacific. For the first time in two years rain fell, and the barren desert was suddenly transformed, in Wells's words, into 'splendid pastoral country'.[82] After a few more months of exploration the party finally made it back to Adelaide. An official inquiry into the troubles of the expedition later concluded it was 'a sorry affair out of which no one emerges with undisputed credit except Lawrence Wells . . .'[83] Sadly for Lindsay, the cruel timing of the El Niño had stolen his glory. And it was not the last time that proper recognition eluded him. In passing through the Coolgardie region during the expedition, Lindsay noted in his journal and maps that the area was 'undoubtedly auriferous'[84] – or gold-bearing – a full year before A.W. Bayley's discovery in the same spot triggered the West Australian gold rush and brought Bayley lasting fame.

1941

Only the exceptional El Niños of 1982–3 and 1997–8 exceeded the strength of the 1939–41 event last century. Unusually for an event that stretched over two years, the peak occurred in the second year, after a relaxation in late 1940 and early 1941. Stories of its severity stretch around the globe.

In war-torn northern China, the deficit of rain was said to be 'a heavier blow' than the Japanese mopping-up campaign. Preoccupied with fighting the war, Chiang Kai-shek's government provided lamentably little assistance, but at least at one point it did manage to smuggle 50,000 bushels of grain across

Japanese lines into the mountains of Honan and Shenxi. The communists earned a far superior reputation for famine relief, and at least one million refugees crossed the Yellow River north into the communist-stronghold border region, where attempts to provide food and land were made, despite the region's own crop failures. 'The bark was stripped from every tree so that the trunks presented a strange white appearance like people stripped of clothes', observed the travelling American journalist Jack Belden in the spring of 1942. 'Women exchanged their babies saying: "you eat mine, I'll eat yours".' When a man was going to die, he dug a pit and sat inside and asked neighbours to fill in the earth when he was dead. Afterward, however, no-one could be found to fill in the pits for all were either dead or too weak to shovel earth.'

By 1941, excessive rains lashed the west coast of the USA, and were repeated in February 1942 when the Sacramento River had its third worst flood last century. Coastal Peru was drenched by flooding rains too, while the warm waters off the coast left 'millions of dead fish floating in the bay, and others cast up on the beach'. In Kenya, heavy rains caused an outbreak of potato blight after a shipment of infected seed potatoes was sent from England to grow extra supplies to feed the allied troops. Drought and a long sequence of frosts in Papua New Guinea led to devastating famine. In Europe the harsh winter of 1941–2 reached epic proportions, the worst ever recorded in the Netherlands, Romania, Sweden and Finland.

However, it was in Russia that the unusual weather of 1941 really impacted on the course of twentieth-century history.[85] By July, Hitler's *Wehrmacht* stood poised on their eastern front, ready to invade Russia and capture Moscow.[86] In what was debatedly the most significant mistake of World War II, Hitler spontaneously decided to first redirect his efforts to the destruction of Leningrad in the north, and the capture of the

Caucasus oil fields in the south. For three vital months, the attack on Moscow was delayed. Even so, the German army retained more than enough men and fighting power, provided the winter was kind to them.

Hitler never expected he would still be fighting on his eastern front by the end of the year. A gross misjudgement over the character and strength of the opposing forces, and an ignorance of the potential severity of Russian winters, led him to believe that resistance would soon be broken. Little preparation had been made to equip the troops with the clothing and food needed to survive a harsh winter, nor to secure the armaments, tanks and motorised vehicles needed to operate in extreme temperatures.

Mud was the first hazard to strike the *Wehrmacht*. Autumn was regularly the season of mud, but the exceptional rains of October made the roads impassable. Virtually all transport, from tanks to horsedrawn vehicles, became bogged, immobilising the entire German army for a month. The attack on Moscow was stalled until the winter.

The forecast for the German army, issued by the meteorologist Franz Baur in October, was for a mild winter. Baur's prediction was based on the fact that never in recorded climatic history were more than two severe winters experienced in succession, and the winters of 1939–40 and 1940–1 had already been excessively cold.

By the time the first cold wave hit in November, much of the German equipment had ceased functioning, and demoralised troops were stripping the local villagers of their clothing. By the end of November the average daytime temperature was below –20°C and continuing to fall. After the war, a Russian meteorologist described it as 'this coldest winter of the last 200 years in northern and eastern Europe'.[87] When the Russians launched a ferocious counter-offensive in early December, the Germans struggled to retreat, abandoning frozen soldiers and much equipment. That winter, between 100,000 and 110,000 men died of frostbite or frost related injuries, a casualty rate that during its December peak even exceeded that resulting from enemy action. Hitler rejected personal responsibility for the humiliation, instead attributing blame to the 'surprisingly early outbreak of a severe winter in the East.'[88] A German radio broadcast also blamed defeat on the weather, making no mention of the Russian army's triumph.

The Russian counter-offensive marked the beginning of the end for the Germans. Hitler had missed the window of opportunity to take Moscow and although the Germans renewed their offensive in the summer of 1942, they never fully recovered from the losses sustained in the 1941–2 winter.

In December 1941, Hitler was shown a diagram of the climatic data of 1941 and 1812, highlighting the years' remarkably similar conditions. Known for his sensitivity over the failure of the *Grande Armée*, Hitler is reported to have exploded: 'Those damned meteorologists, they are also talking of Napoleon.'[89] He was not to know that they were also talking of El Niño.

6

The Price to Pay
Counting the Cost in the Modern World

'My father used to have over 100 head of cattle. We were a rich
family and a happy family and had many friends. We used to cross
the Sahel with our animals going from one waterhole to another
. . . But the drought came. At first we did not pay much attention
to it because where we live it is always hot and dry and does not
rain much. But last year was not like other years. This time there
were no rains at all. First the grass dried up, then the waterholes.
Then our cattle started to die. . . . My father was told by his
brother that cattle and sheep and even camels were dying everywhere in the Sahel and that soon people would begin to die too
because there was no longer any food for them to eat. He told us
that everyone was going south to Niger because there we would
find food and water for ourselves and our animals . . . My father
decided that we must go . . . My mother was too weak to go so we
left her and my two young sisters behind in Gao. They were very
sick . . . My father died one day after we got there [Niamey]. I do
not know whether my mother and sisters are still alive'.

Interview with Tuareg boy from Mali, 1973.[1]

The sixty odd passengers and crew aboard the Barito River
ferry never saw the tug that hit them on the morning of 26
October 1997. Even after the sound of the beam-splintering

crash reverberated around the boat, it was impossible to see what had caused the noise. The market ferry had been easing its way through a dense white shroud of smog that had descended over the river, reducing visibility to three to four feet.

In normal circumstances, everyone might have survived. The boat was close to the bank, and it would have taken minutes to sink. But the impenetrable smog had not just blocked out the sun, it had also smothered any sounds of outside life. In the disorientating void, few on board knew where the nearby bank lay. And no one on the bank could see or hear that a disaster was taking place just swimming distance away. In the panic, twenty-eight people drowned.

On Sumatra, only a month beforehand, the smog helped claim an even greater death toll. Flying in through thick haze, thirty minutes before landing, a Garuda Airbus began to make its descent towards Medan Airport. At this point, according to an unofficial transcript of the conversation between the cockpit and On Route Control, the pilot was asked by Control to make an unexpected turning to the left. To Captain Hance Rahmowiyogo, with twenty years' experience and over 12,000 flying hours under his belt, this did not sound right. But the thick haze prevented him from making his own assessment. 'Confirm are we cleared from a . . . mountainous area,'[2] he questioned. The response was affirmative. Four minutes later the captain was still questioning his instructions from Control. Moments before the impact as the mountainside appeared suddenly through the haze, the pilot uttered his last words: 'Allahu akbar.' All 234 aboard were killed.

For a full six months from July 1997, large parts of Indonesia were enveloped by a choking pea-souper, which stretched across about 100 million square kilometres from southern Thailand to the Philippines, at times even reaching northern Australia. Singapore and Malaysia were also badly affected. A potent combination of ignorance, corruption, misguided land policies, and one of the two biggest El Niños of the last century had all conspired to create a political, social and economic disaster.

Fire is a traditional tool of the slash-and-burn agriculture of Indonesia's many forest-dwelling peoples, and every year hundreds of small fires are lit to clear the way for their small plots. Normally, most fires will not burn for longer than a few days, unable to travel into the wet rainforests or withstand the short spells of rain that douse them. In El Niños, the situation is different. Normally wet dipterocarp forests become flammable enough to burn only after prolonged drought. Historically infamous fires, such as those in 1877–8 and 1914, have invariably been connected with El Niño years. In fact, the great fire of 1983, when 3.2 million hectares were burnt in east Kalimantan and coal seams burnt for several years, is recognised as the single largest ever recorded. The 1997 fires were even larger, although more scattered – covering an estimated five million hectares. But what really set the 1997 blaze apart from the rest was the smoke and haze it generated. This time, fires in Kalimantan and Sumatra were allowed to burn into the freshwater swamps, which had dried out considerably in the drought. Peat swamps contain vast stores of carbon that release much more smoke than normal forest fires. One analyst wrote at the time: 'Whereas the impact of fires concerns mainly foresters and conservationists, it is the smoke that causes politicians and economists to react.'[3]

While El Niño can reasonably be blamed for the extent of the fires, it certainly cannot be attributed to the reasons they were lit. Although fingers were initially pointed at the small-scale slash-and-burn farmers, later, more authoritative estimates found at least 80 per cent were started by plantation owners, industrial estates and land-clearing projects for the transmigration of Javanese farmers. Officially the government prohibits the lighting of fires, but such laws are weakly enforced, and a blind eye is turned to most illegal logging and clearing practices. In places, fires were deliberately used to cover illegal logging, and to blur concession boundaries. In fact, many of the government's own land-use policies have actively encouraged land clearance. In 1995, President Suharto announced that in order to achieve self-sufficiency in rice and to ease the population burden of Java, a million hectares of peat bogs in central Kalimantan were to be converted into paddies for transmigrants by 2000. Since then, gigantic canals have been dug to drain the swampland and provide irrigation, leaving remnant patches more prone to drought. A further 2.7 million hectares have been earmarked for oil-palm plantations, almost double the existing area, as part of a policy to develop a cash-crop economy. Companies intent on meeting these targets as soon as possible have not been fussy about the means of achieving them. As the price of timber supplied to processing mills has been kept artificially low by the government, it has generally proved cheaper to burn the forest than clear it. Some individuals start fires too, burning plantation trees in retaliation for being driven off their land.

The direct human cost of the fires and smoke is difficult to quantify. The haze that built up over Indonesia resulted from the accumulation of suspended particles and chemicals from the fires, including small particles of damaging soot as well as various toxic gases. The heaviest air pollution was felt by eastern Borneo. At one point the pollution index at Pontianak in west Kalimantan reached 1,890 μg/m^3, a level considered 'highly dangerous', and nearly forty times the level considered 'normal' according to the World Health Organisation. Towns above 300 μg/m^3 and visibility below 100 metres were considered for evacuation, although no evacuations actually took place. At one school in Jambi, it was reported that children were roped together in the school grounds to stop them getting lost. Over the whole of Indonesia, at least a million people were estimated to have experienced eye, skin and respiratory problems, and around 40,000 had to be hospitalised. Drinking water was affected too, particularly after the short bursts of rain that were highly acidified. And besides the ferry sinking on the Barito River, and the Garuda flight crash near Medan, there were several other significant transport disasters blamed on the smog, including two ships colliding in the Straits of Malacca, killing twenty-nine people.

There were also some moderate losses suffered by the islands' flora and fauna. Although most fires initially began in scrub and degraded forest, which are not generally important for wildlife, many soon spread into the prime wildlife habitat of mature rainforest, including eleven national parks. In particular, Tanjung Puting National Park, an important reservoir of orangutans, was badly affected. Across Borneo,[4] an estimated 5000 of the apes died, some by being pushed into the open where they were killed by villagers seeking their young as pets. It can also be assumed that many waterbirds suffered, as the peat swamps are important as seasonal wetlands.

Although many farmers were seriously hampered by the smoke, it was the drought that caused the greatest hardship, particularly in parts of Java, Irian Jaya and east Kalimantan. Many villagers were forced to revert to a diet of cassava and other root staples. The rice crop totally failed in some parts, and production for 1998 declined by 8–10 per cent in the country as a whole. In Irian Jaya, villagers living above 2,200 metres suffered badly when a prolonged series of frosts killed such garden vegetables as sweet potato and taro, and most tree crops, leaving only pig fodder to be salvaged. Over 400 highlanders died of starvation, and over 400,000 were significantly affected by food or water shortages.

The cost to the region's economy was enormous. Indonesia alone lost £300 million in crops, £310 million in timber and £1,100 million in direct and indirect forest benefits. Significant costs were not confined to Indonesia. Tourist bookings fell by a third in South-East Asia, resulting in £160 million losses for the tourism industry. The total economic costs were estimated at £2.8 billion, Indonesia bearing 85 per cent of that, equating to 2.5 per cent of its Gross National Product. This figure does not include reduced crop productivity, long-term health damage or loss of investor confidence.

The drought and associated smog could not have come at a worse time for President Suharto. He was already presiding over an economy that was reeling from the crisis of the Asian stock market. As a result of the drought, and devaluation of the rupiah, rural workers flooded into the cities, compounding the unemployment problem. Prices for common goods rose sharply, and student opposition was building. His health was also deteriorating, and questions were being asked in the press over his increasingly erratic behaviour. Responsibility for the policies that led to the fires was not accepted by Suharto, and the blame for them, at least at first, was publicly attributed to El Niño. Even when 186 companies were later publicly named, less than thirty had their licences revoked. Relations with Indonesia's important ally, Malaysia, suffered too, not least because of the adverse publicity that the country received over the pollution fears leading up to the 1997 Commonwealth Games in Kuala Lumpur. The arrival of the smog soon became a very visible and persistent symbol of the cloud of mismanagement and corruption that had settled over the nation. The conflagration in the forests was soon matched by uprisings in the cities, and Suharto was forced to retire in disgrace in May 1998.

Indonesia's experiences of the 1997–8 drought say a lot about the modern nature of El Niño-related natural disasters. El

Niño events affect modern society in very different ways from the past. The costs in terms of human life have been much reduced, and seldom do the great famines of the past occur. Today a range of factors such as famine relief, increased wealth, improved medicine and flood prevention schemes serve to insulate those closest to the climatic impacts. However, as the risks are spread further afield, involving foreign governments, markets and industries, far more people are ultimately exposed to El Niño's vagaries. While figures running into billions of pounds are often cited in relation to El Niño's global damage bill, they say more about El Niño's specific effect on the United States than they do about the huge social costs experienced in the developing world.

Quantifying the global cost of an El Niño event is almost impossible to do in any meaningful way. Firstly, establishing a causal link between climatically anomalous events and El Niño is extremely difficult. Droughts, tropical storms, floods, tornadoes and frosts can occur in any year, and even when they do coincide with El Niños, it is very hard to prove that they would not have happened otherwise – particularly in the case of single weather events. Maybe the firmest conclusions can be drawn from long-term analysis. One study, limited to an analysis of a thirty-year period, concluded that natural disasters are twice as frequent in the calendar year following the onset of an El Niño as they are in other years. The greatest correlation was found in southern Asia and southern Africa. Another study comparing the numbers of people affected by natural disasters in El Niño and non-El Niño years found that 2.7 per cent more of the world's population suffered during the calendar years in which El Niños were recorded.

However, there are certain disasters which, because of their unusual location, the prolonged nature of the anomaly or the way that they fit the long-term pattern of El Niño 'teleconnections', can be reasonably attributed to El Niño. The floods of the Mississippi River in 1993, for example, resulted from highly unusual weather conditions that persisted for several months. Most of the rain was caused by the polar jet stream that was dragged further south than usual, bringing cold Canadian air masses from the north into extended contact with moist air from the Gulf. Rains had already saturated the upper catchment area of the river during winter and autumn, so that when flooding gradually escalated in April, the tributaries were so full that the water could not be channelled.[5] It was the biggest US flood in 133 years of records, with only the flood of the 1844 El Niño coming close. The Army Corps of Engineers reported that no other natural disaster affected so many for so long in the country's history. The total damage bill spread across twelve states was estimated at US$18 billion. Fifty-two lives were lost, 1,000 levees collapsed, 50,000 homes in 75 towns were completely submerged, and road and cross-river rail transport was totally paralysed. A newspaper survey found that 18 per cent of people believed the deluge to be 'God's judgement on the US people for their sins'.[6]

In February 1983, Australians experienced two memorable events in the space of eight days. The country had been gripped by drought since April 1982, with the lowest rainfall on record and searing temperatures throughout the summer. The drought conditions affected 60 per cent of the nation's agricultural and pastoral properties. Grain production tumbled to 58 per cent of the average, 100,000 jobs were lost and the total cost to the economy was estimated at A$7.5 billion.

The first El Niño-related disaster struck the city of Melbourne in an unusual disguise. Most of the city's two and a half million inhabitants probably knew of the erosion problem in the Mallee, the semi-arid region about 500 kilometres northwest of the city. However, few would have had any real sense of the magnitude of the problem. Few, that is, before 8 February. Ever since farmers imported European agricultural methods to the Mallee, the topsoil has been blowing away. Minor dust storms have long been a feature of the region, but usually the soil never travels further than a few miles, falling where the winds die and piling up against fencelines or roadside windbreaks. But in 1983, strong, unabating 50km/hr winds carried the topsoil all the way to the big smoke. Like a slowly unravelling red carpet, the storm smothered the city in a cloud of choking dust. By the time it had passed, 250,000 tonnes of topsoil had been dumped on Melbourne, with an average 20 kg coating each suburban block.

Only a few days later, El Niño revealed a more familiar face to a swathe of the country's south-east coastal region. On the morning of the 16th, firefighters woke to ominous temperatures of 35°C, peaking later in the day at 43°C. Humidity was low and the air exceptionally dry. The forest fuel layer was so inflammable that forest authorities labelled conditions 'super critical'. However, firefighters hoped the warm air brought by the 60km/hr northerly winds from the interior would be relieved by the south-westerly wind change expected later that day. Meanwhile fires broke out in several places along the south-east coastline, including bush fires around Melbourne and Adelaide. The cold front finally arrived in the afternoon but instead of bringing welcome relief in the form of rain, it only served to swing the winds around 90 degrees and open up the fire, turning the elongated flanks into fronts. Sparks were fanned well ahead of the blaze, up to 40 kms in places, and even people fleeing in vehicles found their escape routes blocked by flames. In all the fire claimed seventy-five lives and injured a further 3,500 people. Some 1,700 homes were destroyed, 335,000 hectares of land burnt and 250,000 sheep lost.

At the peak of the drought, with Australia still in pessimistic mood, the government foolishly called a general election.

Before rains brought an improvement in conditions, it was voted out of office.

There are several ways in which El Niño's impacts are felt by human societies. The first and most obvious is its effect on food production. Statistical correlations between the El Niño cycle and yield fluctuations have been made for several crops, including wheat and sorghum in Australia, corn in the United States and sugar cane in Cuba. The strongest connection of all has been found in Zimbabwe where 60 per cent of the variance in maize yield can be accounted for by sea-surface temperatures in the eastern tropical Pacific. There is much anecdotal evidence of similar links with other crops, such as cotton in Peru and rice in Indonesia, and it is likely that future analyses will uncover statistical relationships with many more.

The social and economic cost to a nation as a whole depends heavily on the degree of development and the extent to which the nation relies on its agricultural sector. Australia's great drought of 1982–3 claimed 100,000 jobs through loss of grain, caused average farm incomes to decline by 24 per cent, saw rural exports tumble by A$500 million, and cost the country as a whole an estimated A$7,500 million. Yet, not a single life was known to be lost as a direct result. In contrast, developing nations that rely heavily on their primary productivity for food security, employment and foreign revenue, are much more vulnerable. The drought that hit Papua New Guinea in late 1997 caused the nationally important Ok Tedi mine to shut down for seven months because the barges for copper concentrate could not move downriver. Production at two-thirds of the country's other major mines was also drastically reduced because of water supply problems. As 85 per cent of Papua New Guinea's export earnings come from copper, gold, oil and silver, this had a major adverse effect on the entire economy.

The continuing ability of El Niño years to affect food

production across vast areas was brought home sharply in southern Africa during 1991–2. Widely considered as the region's worst drought in the twentieth century, every country south of Zaire was severely affected. Critically, El Niño's strongest impacts were felt during the summer rainy season, with virtually no rain falling during January and February, the two most important months for the maize harvest. Cereal production was cut by a half, and 12 million tonnes of grain were imported to make up the shortfall. Although over 100 million people were affected – 40 million directly – a human tragedy was widely averted thanks to a successful international famine relief programme.

An even better global example of El Niño's capacity to affect vast areas simultaneously dates back to the 1972–3 event when several important grain-producing regions, including Russia, India, China, Australia and Indonesia, were all hit. With a switch to soybean by US grain producers, in response to the El Niño driven fall in Peruvian fishmeal, grain prices soared and for the first time in twenty years, global food reserves fell.

El Niño's effect on global commodity prices is generally buffered by the ability of some countries to make up part, or all, of another's shortfall. In fact, the contrasting fortunes experienced by different countries during an El Niño year may even assist this. At the same time that drought is damaging Indonesia's coffee output for example, there is a likelihood that heavier rains in other coffee producing countries such as Kenya and Vietnam will produce bumper crops. Wheat is another crop that is affected independently in several countries at once, but its price will be most dependent upon how El Niño affects the harvest of the US Midwest, producer of 90 per cent of the world's grain.

Perhaps the least known of El Niño's impacts on people is its influence on the risk of disease transmission. It has long been

accepted that outbreaks of disease have been historically
linked to El Niño related famines, often occurring several
months after the weather anomaly. However, it was believed
that this was the result of famine weakening a population and
reducing its natural resistance to disease. In the light of
modern epidemiological studies, it can be shown that the
same climatic factors that contributed to a famine might also
have affected the transmission of disease.

Every pathogen has a natural ecological boundary, be it
determined by rainfall, temperature or humidity, within
which the pathogen can survive and remain infective. Where
the pathogen is transmitted by a carrier, or vector, the
boundary may also be limited by the ecological range of
the vector. Where a geographic region remains consistently
within those boundaries, resulting in a high rate of endemic
infection, a degree of protective immunity generally emerges.
But on the margins, particularly in areas where the disease is
active only periodically, exposure to infectious bites is reduced and immunity is gradually lost. It is often in these areas
– just beyond the boundary of the disease's natural range –
that a small change in climatic conditions can affect millions
of people.

To date, epidemics of malaria, the greatest scourge of all
infectious diseases, have shown the greatest correlation to the
El Niño cycle. Malaria remains a major health problem and
an obstacle to economic development for many tropical
countries. Between 3 and 5 million new cases are recorded
each year, with 1.5–3 million people dying, most of them
children. It is projected that in another seventy-five years at
the current trend of global warming, the percentage of the
world's population living within a potential transmission
zone will increase from 45 to 60 per cent.

In 1908 the first symptom that all was not well in Lahore
was the total disorganisation of the train service. Not

renowned for its efficiency at the best of times, stations were suddenly closed and trains cancelled, owing to the sheer numbers of staff falling sick. It soon dawned on the city's inhabitants that it was not just Lahore but the whole of the Punjab that was affected. In Amritsar, home to around 160,000 people at the time, the physician Major Samuel Christophers reported that 'almost the entire population was prostrated and the ordinary business of the city interrupted'.[7] For weeks food was difficult to obtain because of the lack of street vendors, and labour was unprocurable. Over the two peak months of October and November, the death toll exceeded 300,000, or as expressed in the colonial language of British doctors, equated in places to 'five hundred per mile'.

Bouts of epidemic malaria had long been the scourge of the Punjab, yet the degree of mortality experienced in 1908 had not been matched for several decades. And neither had the rainfall. Precipitation was exceptionally high during the La Niña of 1908. A connection between malaria and rainfall was commonly suspected, but until the Punjab epidemic one had not been scientifically identified. For the first time in epidemiological history, a statistical relationship was established between disease and climate. And one of the two people to do it was the same man credited with the earliest study of El Niño itself – Gilbert Walker. Assisting the work of Christophers, Walker helped establish that a 'high degree of correlation exists between the monsoon rainfall and the incidence of autumnal malaria'.[8] Importantly, it was also noted that epidemics tended to follow several years of unfavourable rainfall.

Much of the Punjab, today split between India and Pakistan, is semi-arid, alluvial plains with many areas receiving average annual rainfall of less than 25 centimetres. In most years, mosquitoes are unable to find enough suitable pools of water in which to breed. In the years of heavy rain and flooding that tend to occur in La Niñas, as in 1908, the availability of suitable breeding sites allows an exponential growth in mosquito numbers. At the same time the increased humidity permits a longer lifespan. Mosquito longevity is essential to malaria transmission because the host needs to live sufficiently long for the parasite to complete its sexual stage, and therefore to pass on the infection. It is thought that epidemics in desert fringe zones of the Sahara, southern Africa and western South America are affected by La Niña and El Niño events in a similar way.

In other parts of the world, the mechanism by which an El Niño year may trigger an epidemic is somewhat different. In some wet zones such as Sri Lanka, malaria is restricted by too much rain. So when the south-west monsoon fails, as it tends to do during an El Niño, normally flowing streams turn into stagnant pools conducive to mosquito breeding. In such years, malaria transmission increases four-fold on average. In other normally wet, humid regions such as Venezuela and Colombia, where El Niño years tend to be drier, epidemics are strongly associated with the year after an El Niño. The reason for this is still unclear, but a drought may act to reduce the number of biting cases, or it may reduce the numbers of mosquito predators. In the higher rainfall of the subsequent year, the population immunity may be lowered, and the recovery of predator numbers may lag behind the reappearance of the mosquito.

In the tropical highlands and higher latitudes, the natural range of the malarial pathogen is limited by temperature. Below a certain threshold, sexual development in the parasite is halted. In the malarial fringe areas of the Himalayas, the warmer autumn and winter months that tend to be associated with El Niño years increase the altitude of the parasite's natural boundary. Additionally, the warmer temperatures

decrease the time lapse between generations, as well as shortening the incubation period for the parasite, acting together to
allow more mosquitoes to become infectious quicker. In
north-west Pakistan, malaria epidemics in recent years have
worsened with the influx of three million Afghan refugees
with even less immunity than the local population. In Irian
Jaya in 1997, malarial infective mosquitoes spread up the
valleys to highland communities not normally exposed to the
disease, causing infection rates of over 90 per cent, and
exceptionally high mortality rates of up to 20 per cent.
Estimates of the death toll range between 1,000 and 6,000.

In other parts of the world, El Niño's effect on the transmission of dengue fever is an even greater scourge. Also
known as breakbone disease, dengue can be severely incapa-
citating, causing headaches, fever and painful joints, and
sometimes even death. With the disease having undergone
a major resurgence in recent decades, it now affects around
2.5 billion people, making it the world's most common
mosquito-borne viral disease. Of most concern is the emergence of the new, more dangerous form known as dengue
haemorrhagic fever, which inflicts a 5–10 per cent mortality
rate. Dengue's principal carrier is the domesticated mosquito
Aedes aegypti, which thrives in artificial containers such as in
tyres and debris that are common in urban areas. The
mosquito's range is currently limited to the tropics between
30°N and 20°S, because sustained cold temperatures outside
that range kill adults and over-wintering eggs. Outbreaks can
therefore occur in years in which the minimum temperature is
increased. Furthermore, unusually hot, humid conditions also
act to cause smaller larval size, resulting in smaller adults,
which in turn have to feed more frequently to produce an egg
batch. As a consequence, the incidence of bites increases.
Moderately strong correlations with dengue epidemics have
been shown for El Niño and La Niña years in several

countries across southern Asia and in many of the Pacific Islands.

In East Africa, El Niño's heavy rains are associated with outbreaks of Rift Valley fever (RVF). In early 1998, the region's first outbreak of RVF in a very long time followed hot on the heels of the worst rains for thirty years in north-eastern Kenya and southern Somalia. Mosquito eggs lying dormant in the grassland depressions came to life with the arrival of floods. Cattle losses soared to 70 per cent in some herds as a result of RVF and foot rot caused by standing around in water. One refugee camp, home to people forced to evacuate their flooded homelands, was actually dubbed 'El Niño' by the locals, following reception of a BBC World Service broadcast explaining the origins of the floods.

In Australia, El Niño years have been shown to increase the incidence of both Ross River virus and Murray Valley en-cephalitis (MVE) epidemics. Ross River virus leads to epi-demic polyarthritis, a very painful affliction of the joints and muscles. In parts of the country, virus activity is thought to be correlated to La Niña's heavy rains. In arid country, the virus may remain alive for many years in dormant mosquito eggs that only hatch out as infected adults in response to rainfall. Although less common than the virus, MVE can be much more severe, with a third of cases dying and the same proportion suffering brain damage and personality disorders. A strong association between MVE outbreaks in the south-east of the continent and La Niña years has been noted. In these years, heavy summer and autumn rainfall leads to high numbers of the mosquito vector.

Elsewhere other vector-borne epidemics have also been associated with the El Niño cycle. Among these are Japanese encephalitis in India, West Nile fever in southern Africa and equine encephalitis in the Americas. Horses have also died in huge numbers from African horse sickness in southern Africa, sometimes up to 300,000 in a single epidemic.

Besides the effect on the host dynamics of vector-borne diseases, the heavier rains of El Niño and La Niña years have also been associated with water-borne diseases such as cholera, typhoid, hepatitis A and E, and shigella dysentery. The El Niño connection with cholera in particular has been the focus of much attention. It has recently been shown that cholera can survive in marine algae. A linkage has been made between the incidence of algal blooms and sea-surface temperatures with cholera cases. It is therefore suspected that the sea-surface temperature changes associated with El Niño years may be a factor in causing outbreaks far from their normal range. The unusually warm water conditions during 1991 may have been a factor in the first cholera outbreak in Peru for over a century. Within months it had spread to all South American countries except Paraguay, causing half a million reported cases and 5,000 deaths. In Peru, more than 1.5 per cent of the population was struck.

By the 1970s, with the success of eradicating smallpox, and with tuberculosis and polio on the decline, many epidemiologists were becoming increasingly confident that the battle against infectious disease was gradually being won. In the last two decades of the twentieth century, that view had to be radically revised with the emergence of over thirty new diseases, including HIV, Legionnaire's disease, Lyme disease, Ebola virus, and toxic *E. Coli*, as well as the resurgence of old scourges such as malaria, plague, tuberculosis, cholera, dengue and yellow fever. The reasons for this trend are very complex, but climate change and climatic variability have played their part. In at least one case, the disease has been shown to have been a natural infection of an animal species that crossed over to human populations during a climatic extreme.

The first time anyone rang alarm bells over the appearance of a mysterious killer diesease in south-western America in 1993 came after the death of a Navajo long-distance runner. To the astute New Mexico physician, Bruce Tempest, it was not just the death of a fit, young person to sudden pulmonary failure that was intriguing, but the fact that he had just attended the funeral of his fiancée, who had died in exactly the same manner. After making enquiries, Tempest quickly identified several similar cases in the region. Laboratory tests narrowed the culprit down to an insidious new form of hantavirus.

The virus had first come to prominence during the Korean War, when 3,000 GIs were inexplicably struck down with haemorrhagic fever, many within the vicinity of the Hantaan River from which the disease takes its name. Further outbreaks appeared in subsequent years across Asia and Europe, but instead of infecting the kidneys as the earlier strain had done, the new US strain infected the lungs, eventually causing the capillaries to leak. Most importantly, the mortality rate of the new strain was over 50 per cent.

Why the outbreak occurred when it did might have remained a mystery had the mammalogist Robert Parmenter not been conducting a study on rodent populations in New Mexico. In the summer of 1993, right before Parmenter's eyes, the ecological balance suddenly changed. All of a sudden, deer mice were everywhere. Six years of below average rainfall had already reduced the populations of the mice's main predators – coyotes, foxes, snakes and owls. Then the arrival of heavy spring rains across south-west USA, a common response during El Niños, produced an abundance of grasshoppers and piñon nuts, the mice's favoured food. As a result, deer mice swelled to ten times their usual density until late summer, when the predator numbers also began rising.

It was already known that hantavirus could be spread by

aerosol from rat urine and faeces, so when Parmenter was asked if he had observed any unusual rodent phenomena, he was able to immediately pinpoint the virus's probable reservoir. Even with the public health warnings that followed, advising people to take special care in keeping deer mice out of their homes, forty-six people were struck down, twenty-six of them fatally. By late summer, the rise in predator numbers was beginning to impact on mice numbers, and the outbreak abated.

The virus was likely to have been present, but dormant, all along. It took special environmental conditions to influence the numbers of host and host predators. Similar outbreaks of the pulmonary form of hantavirus, involving smaller numbers of people but with equally high mortality rates, have occurred subsequently throughout the Americas, from Argentina to Canada.

7

Troubled Waters
Impacts on the Marine Environment

'. . . the fishermen on the coast began to talk about an undertow in the coastal current which pulled their nets in the wrong direction and brought strange fishes up from the deep. They had to cease fishing due to the numbers of huge sea-lions which destroyed their nets and gathered on the beaches'

Thor Heyerdahl[1]

The Peruvian anchoveta is a small, nondescript fish. Not to be confused with the common anchovy familiar to pizza lovers, the humble anchoveta, *Engraulis ringens*, holds enormous significance in the Peruvian marine ecosystem. Compacted into dense schools of a silvery mass that swirl and dart within a thin band of cold coastal current, the total anchoveta population consists in normal years of an estimated 10^{12} individuals – about the same order of magnitude as there are stars in the Milky Way. Its abundance is an indication of the extraordinary productivity of the Peruvian coastal ecosystem, and the biomass that anchoveta support is integral to the normal functioning of the entire food web. The corollary is that when numbers crash in El Niño years, the whole ecosystem is disrupted.

The lives of Peruvian fishermen have long been disrupted by El Niño's periodic visitations. The productivity of the richest fishery in the world nosedives as broadscale processes across the whole of the Pacific basin distort the normal currents and water temperature off the Peruvian coast. For months on end, fishermen are forced to seek alternative employment. In the past, many subsistence fishermen abandoned their normal livelihoods altogether and headed inland where deserts blooming from rare downpours briefly offered an alternative source of food. Nowadays, many snatch the opportunity for net and boat repairs, and fall back on their savings. Some continue to fish, seeking out the tropical species that arrive with the influx of warmer waters. But the total volume of the tropical visitors is minute compared with that of the normal assemblage of species, and in addition they are often difficult to market, requiring many more hours of work to extract the same return. Before the scientific community began studying El Niño, Peruvian fishermen had no idea that the temporary loss of their local anchoveta catch was part of a much bigger reshuffle of environmental conditions on the global stage. Nor could they possibly know that similarly dynamic changes were being witnessed in seas around the world.

Anchoveta have long been a vital component of the diet of Peru's coastal people. As early as 4000 BC, the density of fish populations close to shore enabled settlement along the barren coastal strip. Such is the richness of this food source that in the 1970s the archaeologist Michael Moseley proposed that it underpinned the rise of civilisation before 2000 BC along the Peruvian coast. This theory was highly radical at a time when it was universally believed that civilisations could only have sprung from the foundations of agriculture.[2] Even during the era of the great Peruvian states that were built on maize and other grains from their irrigated river valleys, the protein from anchoveta continued to be an essential dietary supplement. In some villages today, fishermen can still be found beyond the surf-line using the same technique that has been used for at least 2,000 years to catch this populous fish from their *caballitos* – one-man totora-reed boats.

Indigenous Peruvians have known about the agricultural benefits of anchoveta for centuries. One of the earliest *conquistadores* described the local farmers' practice of planting with each maize seed 'a head or two of the sardines which they take in their nets'.[3] But the fish only began to assume an economic importance to Peru in the early 1800s when chemists discovered the fertilising properties of guano, or sea-bird excrement – the by-product of massive anchoveta consumption in Peru. Particularly rich in nitrogen and phosphorus, guano quickly became a valuable commodity sought by farmers across the US and Europe. And the prime supplier of this new fountain of riches proved to be the seas off Peru, the vast biomass of anchoveta providing a diet for millions of sea-birds, and the relative lack of rain allowing the steady accumulation of guano on the flat, rocky offshore islands.

Even in the days of the guano boom, the impact on the industry during these occasional years of warm currents and exceptional precipitation was noted as millions of guanobirds disappeared and thousands of tonnes of guano were washed into the sea. But serious concern over the economic implications of these abnormal years began to mount in the 1950s, when the anchoveta itself was discovered as an untapped goldmine. Converted into fishmeal, a feed supplement for the burgeoning American poultry market, or into fish oil – an ingredient of margarines, paints and other products – anchoveta immediately became the highly valuable export commodity that it remains today.

By the 1970s, Peru had transformed itself, in terms of fish-catch volume, from an insignificant fishing nation to the number one in the world. The annual catch of anchoveta had risen within two decades from 0.5 to peak at 12 million metric tonnes, representing 25–30 per cent of all foreign export earnings. In fact, catches achieved an all-time peak during the onset of the 1972–3 El Niño, just as the earliest intrusion of warm water drove shoals of anchoveta to seek refuge in the remaining pockets of cold water close to the coast. Over 180,000 tonnes were hauled in a single day. Weeks later, the population had crashed. Bankruptcy threatened half the fleet, and led to the expropriation of the fishing industry by the military government.

But whereas previously populations had soon returned to their former abundance, particularly when assisted by the favourably cool waters of a La Niña year, this time they didn't. In 1973, the annual catch was only two million metric tonnes, a figure that had only doubled by the time it crashed again in the 1976–7 event. The halcyon days enjoyed by the anchoveta industry in the early 1970s have never been repeated, although more effective management of the industry, and the prohibition of fishing during the peak stages of more recent El Niños, have helped the annual catch in normal years to reach around 7–8 million metric tonnes.

But the social cost of such heavy regulation of the fishery is high, and now with each event the piers and sea-fronts are overrun with tens of thousands of unemployed fishermen leisurely killing time. In 1997–8, for example, 20,000 were left port-bound in Paita, representing 70 per cent of the city's workforce. At least now, thanks to scientists, fishermen understand the biological principles that drive their boom-and-bust industry. The processes that underpin the anchoveta's extraordinary numbers cannot be separated from those that drive its spectacular population crashes, and both are inextricably entwined with the El Niño cycle itself.

Despite being nestled in the heart of the tropics only a few degrees south of the equator, the temperature of Peruvian waters ranges from around a cool 22°C in the north of the country to around 10°C in the south. The first records of these unexpectedly temperate conditions were made by the earliest *conquistadores*, who cooled their flasks from the tropical sun by dangling them overboard. Later, when the great naturalist Alexander Humboldt first observed the temperature of the northward-flowing offshore current named after him, he presumed that the current was bringing up chilled waters from the sub-Antarctic. To a small degree, Humboldt was right. However, the vast majority of the cold water does not come from the far southern latitudes, but from below.

Wherever they occur, upwellings of cold sub-surface water are the basis of the world's richest fisheries. It has been estimated that upwellings are on average over 65,000 times more productive in terms of fish yield per unit area than the open ocean. The world's five main upwellings off the coasts of Peru, California, Mauritania, Namibia and Somalia comprise only 0.1 per cent of the ocean's surface area, yet contribute nearly half of the global commercial fish catch. The Peruvian upwelling itself supplies over 20 per cent. Most of that is attributable to the single species of anchoveta which, to put it into some perspective, represents 60 per cent more volume than the total US fish catch.

While the physical processes of an upwelling occur mainly within 50 kilometres of the shoreline, bands of intense biological production may stretch to over 300 kilometres offshore. For the fish, the essential ingredient being drawn up from below is not the cold temperature, but the nutrients that accompany it. Derived from the breakdown of dead organic matter, most nutrients settle at the bottom of the ocean before they can be consumed by phytoplankton. To be biologically useful, these nutrients must be present in the sunlit layer, the zone to which photosynthesis is confined. In clear waters, this may extend down as far as 150 metres, but in more commonly turbulent conditions, sunlight fades markedly with depth, penetrating only a few tens of metres. The reality for much of the world's open oceans is that the two essential factors for high biological production exist in virtually exclusive zones. In upwellings, the two are brought together.

The food chains of upwellings are very distinctive, with relatively few links between top and bottom. In the Peruvian upwelling, the anchoveta feed directly on phytoplankton at the bottom of the trophic web, and in turn provide the main food source for a diverse array of predators at the top, including other fish, mammals and the largest population of oceanic birds in the world. The anchoveta's existence, and the prosperity of much of the food chain, is intimately tied to the physical processes that maintain the cold, nutrient rich waters near the surface.

The secret of the anchoveta's ability to exploit the peculiar conditions of the Peruvian upwelling is its adaptation to life in the fast lane. Three years is the fish's average lifespan, but it takes less than a year for most females to reach spawning age, an uncommonly early age in the lifetime of a fish. Over a female's lifetime, 50,000 eggs on average are laid. Unlike many other fish species, the anchoveta egg contains little yolk, the usual source of energy for a larva until it can find its own food. With very limited powers to swim, anchoveta larvae depend on a rich and immediate source of food. Because of this, the species is restricted to the zone of highest productivity, in effect the thin band of northward flowing coastal current. Typically this is no more than 1–200 kilometres wide, rarely more than 200 metres deep, and with a temperature varying little from 16 to 18°C. But while no other fish is better adapted to exploit this bounty, no other fish is more vulnerable to its disappearance.

The first sign of an impending El Niño experienced by an anchoveta is probably the arrival of the first Kelvin wave.

Having spent six to eight weeks travelling from its origins in the western Pacific, this sub-surface wave heralds months of abnormal conditions for marine life. The ensuing temperature change can be very sudden. The first Kelvin wave of the 1982 event, arriving on 7 October, ratcheted up the temperature within the upwelling by 4°C in one day. In strong events, the overall rise has measured up to 10°C. Although adapted only to a narrow band of water temperature, the fate of the anchoveta population is not ultimately determined by the increase in temperature but by the impact on the circulation of nutrients and associated availability of phytoplankton. Cold water upwelled from below originates mostly from 40 to 80 metres. Even in an El Niño event, this process continues, but because the influx of warm water has depressed the thermocline, the water that is entrained from below is no longer cold and nutrient-rich. In strong events, the effect can amount to a twenty-fold reduction in primary productivity, clearing waters normally murky green with organic matter within days.

The anchoveta's response to a change in water temperature is almost immediate. But while all fish react by moving towards colder water, this does not result in the whole population suffering an identical fate. Those shoals further from shore tend to head for deeper water further down the continental shelf. To avoid the warmer water above, the fish suspend their daily vertical migration, which would take them into shallow water at night. But even here many fish starve, as food and oxygen levels at these depths are much lower than normal. Many of those that do survive head south, becoming a welcome bonus to Chilean fishermen. Other shoals are drawn to the pockets of cold water that remain close to the coast, offering a fleeting boom for Peruvian fishermen. Such behaviour is likely to be an adaptational response to events of weak to moderate strength, where these inshore pockets of cold water tend to last the duration of the event. In

strong events, however, this behaviour proves the kiss of
death. Locked in by land on one side and warm water on
all others, the final pockets of cold water are overridden by
the influx of even more warm water and virtually all adult fish
succumb either to starvation or to thermal shock.

While the story of the anchoveta's episodic demise has
contributed more than any other issue to focus early scientific
attention on the El Niño phenomenon, it is of course only one
small component in a vast catalogue of changes to the
upwelling ecosystem. Throughout normal years, the upwelling ecosystem remains relatively constant in environmental
variables such as temperature, salinity, light, nutrients, and
oxygen levels, and consequently its inhabitants are adapted to
relatively constant conditions. In upwellings, as in all tropical
ecosystems, by far the most radical changes to a species'
environment occur not between seasons but during El Niño's
occasional disturbances. During mild to moderate El Niño
events, such changes are rarely extreme or long lasting. In
these events, the water temperature of the upwelling, for
example, will seldom rise more than 5°C, and the peak
changes may not last longer than a few weeks. Some degree
of disruption to the normal rates of reproduction, growth and
survival inevitably occurs, but the physiological and behavioural responses of most species enable them to cope effectively with moderate and short-term change.

The effect of a strong El Niño is another matter altogether.
Relative to the seasonal variation that is experienced by the
ecosystem over a normal year, the physical changes are both
severe and prolonged. In the 1982–3 event, for example, the
temperature of the upwelling rose by up to 10°C, and temperatures remained elevated for nine months. A strong El
Niño also heralds raised sea-levels, improved oxygenation at
the sea-floor, greater wave action resulting from stormier
weather, lower salinity and increased sedimentation from

the additional river runoff. A few local species may favour these new conditions but the vast majority do not. For those that cannot migrate, a survival struggle ensues, with death by starvation or thermal shock the consequence for many. Survivors face a period of poor growth, low reproductive success and reduced recruitment of juveniles to the adult population. The earliest stages of life – the eggs, larvae and the young – are the most sensitive to the shift in conditions and prone to total loss.

The most visible signs of change are evident at the top of the food chain. All pelagic fish move location, some like sardines *Sardinops sagax* tending to shift south into Chilean waters, and others like Jack mackerel *Trachurus murphyi* moving into deeper waters. Bottom dwelling species such as hake *Merluccius gayi* tend to migrate further down the continental slope where usual conditions of low oxygenation would normally deter them. Wherever the fish go, they tend to elude Peruvian fishing nets.

The usual cast of cold-water adapted species is replaced by a raft of tropical invaders, but never in the same volume as the normal residents. Manta rays, dolphins, flying fish and hammerhead sharks immigrate from offshore and further north. Some warm-water species are of high market value, such as dorado, bonito and skipjack tuna, going some way to compensate the profits lost to fishermen through the disappearance of their normal catch. Tropical sea-birds too, such as the blue-footed booby, waved albatross, swallow-tailed gull, magnificent frigate-bird and least tern, make an unusual appearance as they follow in the wake of their catch.

Local sea-birds, dependent on the anchoveta, are devastated. The Peruvian booby *Sula variegata*, guanay cormorant *Phalacrocorax bougainvillii*, and Peruvian pelican *Pelecanus thagus*, in particular, abandon their nesting grounds and leave the area, most flying south to Chile. The short-term impact on numbers can be catastrophic. In early 1912, one naturalist arrived on the guano islands to find the disappearance of anchoveta had occurred after the pelicans and cormorants had nested. He surveyed millions of nestlings of various ages 'but all of them dead, sun baked mummies'.[4] In the 1957–8 event, for example, in the days before heavy anchoveta fishing began to reduce bird numbers, the population of guano-birds fell from approximately 30 million to just 6 million in a single year. In 1997, a Japanese film team arrived on Chincha Island, expecting to repeat the success of an earlier mission when they found 800,000 nesting birds. On this occasion they could locate only a thousand. In such times, corpses of sea-birds, along with those of mammals, fish and invertebrates, litter beaches in such quantities that they become a major hazard. Back in the 1880s, Chilean troops occupying parts of Peru after their success in the War of the Pacific had to dig a ten-mile trench to bury the stench.[5]

El Niño conditions can also result in unusual contact between sea-birds and people. Cormorants, normally flock feeders, switch to inshore mullet and attack fishing nets. Many are in such a weakened state that they are caught easily by hand before going into a fisherman's pot. Being the least timid of Peru's sea-birds, pelicans congregate around the ports and fish markets, and have even been known to hold up the traffic at times.

The region's mammals are not spared either. Tied to their breeding sites and limited by their diving ability, South American sea lions, *Otaria byronia*, and South American fur seals, *Arctocephalus australis*, are unable to migrate with their prey, or dive to the new depths where their prey have found refuge. They have to seek larger fish instead of their usual anchoveta, and those seals that do not perish can be commonly found along beaches with their shoulders and girdles poking through their skin. In strong events, mothers are forced to extend their feeding trips from a usual two to three days to up to five days, which for pups dependent on their mother's milk can mean almost certain mortality.

Mortality of all marine creatures can reach such proportions that their decaying bodies give off high concentrations of hydrogen sulphide gas capable of blackening ships' hulls – a phenomenon known colloquially as *El Pintor*, the Painter.

By any standards, the changes imposed on Peru's marine ecosystem during 1982–3, which killed billions of fish, millions of sea-birds and thousands of mammals, could be described as catastrophic. But despite the apparent devastation, within a year of El Niño conditions disappearing, most bird and mammal colonies, as well as intertidal and near-shore communities, had recovered a relatively normal appearance. By the following year, the anchoveta and other pelagic fish had returned to something approaching pre-1982 numbers. And by the time the next El Niño arrived in 1986, the normal food web had more or less been re-established, even if changes to the size and age structure of the population were still apparent.

Environmental variability is a fact of life, and within every ecosystem lies an innate resilience to a wide range of natural fluctuations, enabling it to bounce back even from seeming decimation. In the case of the El Niño cycle, the negative effects of one phase of the cycle are balanced over time by the positive effects of the other.

The key to a species' response to an El Niño event is flexibility. Those species with the greatest ability to conserve resources, switch prey or travel to productive refuges elsewhere can ride out a storm without a pronounced impact on their populations. The most dramatic impacts are experienced by those with more rigid modes of behaviour. These species may be tied to a location through lack of mobility, breeding requirements or dietary specialisation, while all around them the normal conditions of prey, water temperature, salinity, turbidity and oxygenation are turned upside-down.

Even as late as the 1970s, the only marine species recognised to be affected by El Niño were those living in waters off the west coast of South America. In recent years, scientists have detected impacts on marine species right around the globe.

One of the most enduring images of El Niño's biological impacts during recent events has been that on coral reefs, once vibrant with kaleidoscopic colour, bleached white as if blanketed in a winter's snow. The coral's colour comes from the pigmentation of the zooxanthellae, the microscopic algae that live symbiotically within the coral tissue. The zooxanthellae produce food for the coral, and the coral in return acts as host. Bleaching is caused by the loss of pigmented algae, exposing the white calciferous skeleton through the transparent coral tissue. Partial bleaching does occur naturally in patches owing to a variety of different stresses and responses, but the severity of stress from recent El Niños has killed reef colonies hundreds of years old across the tropics of the Pacific, Indian and Atlantic Oceans. These unprecedented scenes have coincided with a growing consensus among scientists that reefs are in serious decline through the cumulative stresses of global warming, overfishing, tourism and pollution. The El Niño related bleaching has acted to compound the crisis and set alarm bells ringing about the state of the world's coral reefs.

Most species of coral are highly sensitive to relatively small temperature change, just as they can also be to small shifts in ultra-violet radiation, light, salinity or pollution levels. Corals can generally tolerate short bursts of these stresses, and as long as some reserves of zooxanthellae are retained, the coral

Underwater temperature profiles across the equatorial Pacific before *(top)* and during *(middle)* an El Niño, showing the eastward shift of warm water within the tropics, the rise in sea level in the east, and the levelling in the angle of the thermocline. A satellite image *(below)* of sea surface temperature anomalies shows the location of the warm pool spread across the eastern equatorial Pacific.

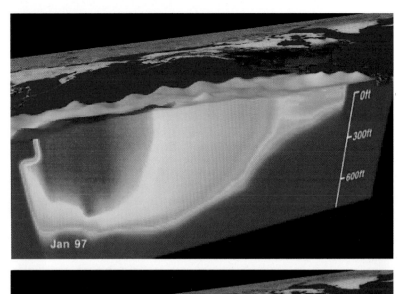

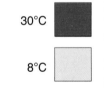

30°C

8°C

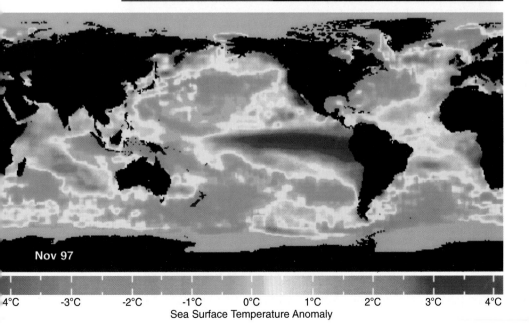

-4°C -3°C -2°C -1°C 0°C 1°C 2°C 3°C 4°C
Sea Surface Temperature Anomaly

'The Discovery of the Potato Blight' by Daniel McDonald, 1852 captures the foreboding experienced at the outbreak of the Irish Potato Famine. Exceptional temperatures and humidity during the summers of 1845 and 1846 allowed the plant disease to spread.

Strong correlations exist between the El Niño cycle and epidemics of vector-borne disease in the tropics and sub-tropics. In the case of malaria, the cycle alters the natural ecological boundaries of the malaria-carrying *Anopheles* mosquito.

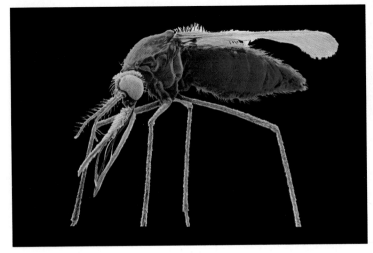

'he Breakaway' by Tom Roberts shows a stockman trying in vain to stop his thirsty
eep stampeding during the Australian drought of 1891.

Above The stricken Exxon Valdez in Prince William Sound, after th vessel ran aground on a reef while diverting around icebergs. La Niñ conditions in 1988–9 had led to a change in the circulation of the Sound's currents, sending icebergs onto the shipping lane.

Left A Bushman woman decked out in beads received through *hxaro*, a system of gift-exchange that forms strong ties between remotely-located families in the Kalahari, a region affected by unpredictable climate.

ove Massive
oding in northern
nya and southern
malia in late 1997
d early 1998 led
 an outbreak of
ft Valley Fever.
normally heavy
nfall in east
rica is common
ring El Niño
ars.

onths of heavy
in during the 1993
 Niño produced
e worst US floods
 record,
lminating in a
mage bill of $18
lion. *(Right)* A
ellite image of the
ississippi, Missouri
d Illinois rivers in
l flood in August
93.

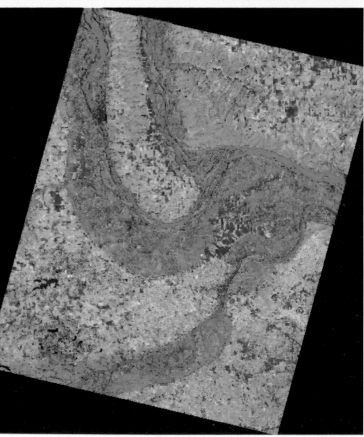

Left The golden toad of Costa Rica has not been seen since the severe drought associated with the 1986–7 El Niño.

Below Exceptionally high numbers of green turtles come ashore to breed on the Great Barrier Reef two years after an El Niño event.

Above Changes to the oceanographic conditions around the Galapagos Islands during El Niño events causes a massive decline in food supply, resulting in widespread mortality in species such as Fur Seals *(left)* and Marine Iguanas *(right)*. In contrast, El Niños usually bring heavier rainfall and improved feeding conditions to the land animals of the archipelago.

Right A healthy Great Frigate chick sits on a nest on Christmas Island in 1984. It is above a dead chick that starved to death the year before. All chicks and ninety per cent of the island's twelve million adult birds died during the 1982–3 El Niño.

Main picture In February 1983, the Australian city of Melbourne was enveloped in a massive dust storm, and eight days later encircled by the country's worst recorded fire (*inset*). Both disasters were associated with the exceptionally severe 1982–3 El Niño.

can bounce back immediately. However, if the stresses are severe or prolonged, all the algae are permanently evicted and the coral dies.

The sources of stress caused directly by an El Niño vary depending on the location. In some regions, the threat may stem from lowered sea-levels which expose reefs to higher radiation and wave action during low tides, or from the lowered salinity due to run-off from extra storm activity. In most cases, El Niño's most lethal weapon is the increase in water temperature resulting from the shift of large bodies of warm water across the oceans. For example, in 1982–3 Pacific waters off Panama, Costa Rica, Colombia and the Galápagos were 2–3°C warmer than normal for several months on end, resulting in up to 99 per cent coral mortality in the shallower depths. Sporadic bleaching has occurred ever since, with the most severe and widespread episodes always coinciding with El Niño years. In 1987, it was the turn of the Caribbean, the Maldives and the Great Barrier Reef. In 1997–8, corals in the shallow zones of many reefs off Sri Lanka, India, the Maldives, Kenya, Tanzania, Seychelles and the Galápagos Islands suffered up to 90 per cent mortality, while at the same time abnormally low temperatures killed reefs in Indonesia.

In recent years, understanding of the bleaching process has undergone a radical transformation and several previous assumptions have had to be revised. It is now known that all corals bleach to some extent annually, and even when bleached white, may still retain some algae in reserve. Talk of a coral's death in some cases may therefore be, as Mark Twain might have it, somewhat exaggerated.

More importantly, some scientists have now suggested that far from being the detrimental phenomenon it might at first appear, bleaching may actually perform an important adaptive function that ensures long-term health. It had been assumed that each species of coral associated exclusively with a particular species of alga. That relationship is now understood to be a much more flexible one, and each coral species can have more than one symbiotic relationship. The particular combination of host and symbiote may exhibit a specific tolerance or preference for environmental variables such as light, depth and temperature. But in response to a change in one of these variables, the host has an opportunity to be repopulated by a different partner. It is likely that if the frequency of stress conditions is altered, a new combination of host and more stress-resistant partner will emerge. Such a theory also explains how coral reefs can be so sensitive to environmental change in the short term yet have survived through extreme change over millions of years.

That is not to say that fears of global warming's ability to significantly deplete the world's reefs are unfounded. The experiences of certain regions during the last two decades of the twentieth century paint a pessimistic picture. How rapidly reefs recover from a bleaching episode depends on several factors. Where mortality has not been high and sufficient corals have survived at lower depths, recovery might occur over five to fifty years, as larvae from undamaged areas recolonise the damaged areas.

In some places such as the eastern Pacific, the recovery process from damage caused since the 1970s might be more in the order of centuries. Here, sexual recruitment of corals is low and recovery is mainly through asexual reproduction in which living corals are fragmented and dispersed by natural processes such as storms, breakages and coral-eaters such as triggerfish. However, the extent of bleaching during an event such as 1982–3 was so comprehensive on some reefs that too few individuals survived to begin the recovery process by asexual reproduction. In such cases, these reefs normally rely on larval dispersal from afar. The nearest reefs upcurrent of the Cocos Islands of Costa Rica, for example, are those of the central Pacific's Line Islands, 6,500 kilometres away.

Even the departure of El Niño's unusually warm temperatures does not mean the process of recovery can begin immediately. Bleached and devoid of much of its defences, a reef is highly vulnerable to further degradation. The immediate problem may be a La Niña event, this time bringing the stress of low tides and unusually cold-water temperatures. Of longer term consequence, however, are the changes to the reef's community composition.

To the naked eye, reefs are rather solid, static constructions. Observed over time however, their shape is constantly being modified. On every reef, the coral structure is being lost not only to intermittent buffeting by wave and storm action, but to the reef's own natural eroders.

In normal circumstances, reefs are gradually accreting material over time, the processes of erosion being outweighed by the rate of deposition from the scleractinian coral. The disturbance of a strong El Niño can upset this balance, providing a temporary advantage to the reef's eroders. After the damage from the 1982–3 warming, the coral framework was being removed faster than it could be replaced. On many of the reefs, live corals were replaced with crust-like sheets of coralline algae that deposit calcium carbonate at much lower rates than corals do. The eroders too, no longer kept in check by the coral's defences, were suddenly presented with an enormous increase in available food, and their numbers and impact escalated. The result was that significant swathes of reef framework in the eastern Pacific receded and collapsed, delaying any recovery of a pre-1982 state, perhaps for centuries.

One study of a Panamanian reef, made shortly after 50 per cent of its coral cover had been lost, found the most visible beneficiary was the Mexican urchin *Diadema mexicanum*.

With much of the coral covered with the urchins' favourite food of coralline algae, urchin densities rose nearly twenty times their pre-El Niño level. Like other herbivorous species that scrape away at the coral as they graze, the urchins can break off large chunks of coral, sometimes toppling colonies in a beaver-like fashion. At such high densities, the urchins were causing immense erosion, several times more than all the other reef's eroders put together, and the reef framework was rapidly receding and collapsing. Yet the threat from the urchin would have been much greater had it not been for another species that emerged from the shadows of the El Niño.

Like other members of its family, the Acapulco damselfish, *Stegastes acapulcoensis*, cultivates its own patch of algal 'lawn', vital to its spawning and feeding.[6] The damselfish weeds out unsuitable algal species and aggressively defends its territory against intruders. Normally, an algal mat cannot grow on the skeleton of live corals, thriving instead on dead-coral surfaces. The mass bleaching of the Panamanian reefs therefore created habitat space for new algal lawns, presenting new opportunities for the damselfish. Although the damselfish population did not actually rise, smaller damselfish previously unable to establish and protect their own lawns moved into the newly available niches.

Should an urchin approach a damselfish lawn, the damselfish reacts with the ferocity of an enraged piranha, either snapping at its spines or picking the urchin up in its teeth and depositing it up to two metres away. In this way, damselfish help protect the reef structure in their territories from serious erosion.

Another species whose fortunes have been known to rise in the aftermath of El Niño disturbance, significantly impeding the process of reef recovery, is the Crown-of-thorns starfish, *Acanthaster planci*. The only organism known to feed entirely on coral, these voracious feeders evert their stomachs over their prey, devouring the coral polyps and leaving the calcareous skeleton without a host. Normally the starfish are sparsely represented among most reef communities, but every so often, outbreaks are triggered that can leave large swathes of dead, white coral in their path.

In the eastern Pacific, many massive corals are protected from starfish attack because the colonies are surrounded by a ring of pocilloporid coral, the main reef-building species in the American Pacific. A host of crabs and shrimps live symbiotically within the branches of pocilloporid corals, feeding on trapped debris and the coral's mucus. In return they guard against predators such as the starfish, attacking them as they attempt to mount the coral colony, the shrimps snapping and pinching at the spines, the crabs levering up the sea-star's arms and clipping at its feet. Additionally, pocilloporid corals also possess their own nematocyst defences. In normal circumstances, starfish will do anything they can to avoid pocilloporid corals, even if it means they are unable to reach their preferred coral prey in the centre of a reef colony. But the pocilloporid corals bleach quickly when exposed to very warm water and when the health of these corals deteriorates, the symbiont guards either move or die, and the starfish are able to breach the reefs' outer defences.

The increase in starfish predation in the eastern Pacific pales into insignificance beside that of the more infamous outbreaks in the western Pacific and Indian Oceans. When the starfish first shot to prominence, its population growth across parts of the Great Barrier Reef was so sudden and widespread that many prophesied the destruction of the entire reef. For several years afterwards, scare stories continued as outbreaks were observed across Polynesia, Japan and Micronesia. Because these outbreaks had not been documented before the 1960s, the initial assumption was that humans had altered the normal balance of reef ecology through pollution or the over-collection of starfish predators. However, these sporadic outbreaks are now thought to be part of a natural cycle.

Increases of five to six orders of magnitude within a single year may amount to millions of individuals, each one capable of devouring up to five square kilometres of coral tissue annually. Once an outbreak occurs, it may run from one to several years before the starfish eat themselves out. With the ability to produce 12 million eggs each month, any condition that prompts even a modest increase in larval survival has the potential to trigger an outbreak.

The primary trigger for these outbreaks is still unknown, but because evidence suggests that they occur synchronously across widely separated reefs, it is likely that some broadscale oceanographic influence is involved. A study of recruitment rates of starfish larvae in Fijian reefs has recently revealed a link with the timing of El Niño events.[7] Years of moderate recruitment were recorded in 1987 and 1994, and especially high ones in 1977, 1983 and 1992 – all El Niño years. In all the other years examined in the study, recruitment was very low.

The response of the red-rimmed thorny oyster, *Spondylus princeps*, to changes that occur during El Niños once accorded it mystical powers. Its white shell is big and spiny and liberally coloured with splashes of purple, orange, red and white, making it one of the most spectacular bivalves in the world. In order to grow such a thick shell and spines, and to produce the cement that binds it to rocks, the spondylid needs to metabolise large quantities of calcium bicarbonate from the water. To achieve this, the spondylid has to live in warm water because the elements required for metabolisation precipitate out of solution in cold water. For this reason, the species' range is limited to the Pacific American coast between Baja California and southern Ecuador.

Valued as jewellery and a form of currency more precious than gold or any other metal, the spondylid shell was sought by civilisations from the Incans of the Andes to the Mayans of Central America. But there were many factors that greatly inhibited supply. The spondylid lives at depths of between 25 and 60 metres, so was only accessible to divers able to free-dive and stay down at these depths for a few minutes. There were very few that could, and these individuals died prematurely. Within the range of South and Central American coastline along which the spondylid can live, there are few suitable reefs on which spondylids can attach themselves and grow.

Supply is also affected by wild fluctuations in the spondylid's abundance in accordance with El Niño's rhythms. It is now suspected that numbers increase substantially with the arrival of warm water in El Niño years, with strong events also enabling the spondylid larvae to disperse further south than usual. In La Niña years, the cold Humboldt current pushes further north, and spondylid numbers fall off rapidly.

Along the coastal desert of Ecuador, El Niño events are strongly correlated with years of excellent rainfall, a phenomenon whose timing related to, presumably in the eyes of shaman-priests, the coincident abundance of the spondylid. The archaeological record has found the use of spondylids in rain-promoting rituals by the Valdivia people of Ecuador's Santa Elena peninsula as far back as 3600 BC. Over two millennia later, the significance of the spondylid had diffused into the highlands, and along the Andes into Peru. By AD 350, the shell was being sacrificed in rituals as far north as Mexico. For the next two millennia, a long-distance trade network thrived, with the demand for shells forming the centre of an exchange system between culturally distinct regions. One vessel encountered by Pizarro's party during the conquest of Peru was laden with silver, gold, jewellery and textiles, all to be bartered for quantities of the shell.

The largest quantities of all were consumed by the Incas. To them, the spondylid was considered 'food' for the gods, and great numbers were sacrificed to induce the gods to grant favourable weather. One chronicle tells of the wrath of Maca Uisa, child of the god of agriculture. After causing great floods and reducing villages to chasms, he scornfully rejects an offering of ordinary food with the words, 'I am not in the habit of eating stuff like this. Bring me thorny oyster shells.'[8] Scenes of spondylid collection depicting divers tethered to balsa-log rafts have been uncovered on Incan textiles, inlaid into wooden carvings and embossed on metal jewellery.

The spondylid's sacred properties have largely been forgotten, usurped after the Christian conquest by other idols such as *El Señor de las Aguas* of coastal Ecuador. However, some of the spondylid's mystical powers have remained in the public memory, and in 1997 the red-rimmed thorny oyster was adopted as the emblem of new peaceful relations between the governments of Peru and Ecuador.

The biological responses in the marine world to El Niño's perturbations are by no means confined to tropical and subtropical waters. Scientists have long suspected that El Niño's fingerprints are also etched on the Antarctic's Southern Ocean, but because of its own internal variability, this impact has been difficult to pinpoint.

Interannual fluctuations in the Southern Ocean ecosystem were first reported in the 1920s aboard the *Discovery*, purpose-built for Robert Scott's first expedition to the Antarctic. The purpose of the *Discovery*'s early expeditions was to investigate British whaling. In the 1920s, still several years before the introduction of the pelagic, deep-water factory ships, the annual success of the port-bound whaling industry was very much at the mercy of the offshore food supply. Based on evidence from whalers, the expedition found that when conditions were relatively cool, blue whales were abundant. Then every few years, perhaps two to three times a decade, it was noticed that the conditions were warmer and the catch was dominated by the smaller, less profitable fin-whales. It was also noted that in a 'fin-whale year' such as 1925–6, a strong El Niño, whales were exceptionally thin and fewer in number.

One suggestion made at the time was that 'fin-whale' years could be linked to oceanographic conditions and an associated shortage of food. It was thought that blue whales, because of their much greater size, would need to travel away from the islands to satiate their food requirements. During these years of blue-whale scarcity, it was also noted that krill, *Euphausia superba*, appeared to be less numerous.

The vast majority of people in the world will never see krill, yet this small translucent crustacean is the most numerous creature on earth. There is an estimated 600–700 million tonnes of krill, nearly ten times the world's total fish catch. Its remarkable abundance provides a key food source for many species, and it can be justifiably credited as being the cornerstone of Southern Ocean ecology.

Much more accurate assessments of krill populations than could be made by the *Discovery* have now shown that krill numbers can fluctuate dramatically from year to year. In 1983, for example, the population was estimated to be only 3 per cent of that of the previous assessment two years earlier. It has also been noted that years of low krill abundance have correlated with years of warmer water and these tend to follow roughly one year after the tropical Pacific has experienced El Niño's warm-water conditions.

The most visible effects of krill shortage have been observed during the breeding season of krill predators when their foraging range is restricted by the need to return frequently to their young. Nowhere has this dilemma been more starkly demonstrated than on the rugged island chain of South Georgia. The island lies just to the south of the Antarctic Polar Front, the natural barrier that defines the limits of the Southern Ocean. Here, two species of albatross – the black-browed and grey-headed *Diomedea melanophrys* and *D. chrysostoma* – as well as the gentoo *Pygoscelis papua* and macaroni *Eudyptes chrysolophus* penguins have all suffered breeding failure when krill is scarce around the islands. In 1994, when krill numbers were down by 90 per cent, the breeding success of the macaroni penguin was reduced by a modest 10 per cent compared to that of a previous study in 1986. This small change reflects the macaroni's ability to switch its prey from krill to amphipods. By contrast, the gentoo penguin, with its more limited foraging range and thinner, dagger shaped beak less suited for catching the smaller amphipods, suffered a 90 per cent decline in breeding success. The 1994 conditions also reduced the breeding success of the grey-headed albatross, whose diet includes squid and lamprey as well as krill, by 50 per cent. The black-browed albatross fared even worse. Its chicks are more intolerant of reduced food supply because of their faster growth rate and shorter fledging period. As a consequence, breeding success plummeted by 90 per cent. The krill-dependent Antarctic fur-seal, *Arctocephalus gazella*, which breeds almost exclusively on South Georgia, lost 30–35 per cent of its pups the same year – two to three times the normal mortality rate.[9]

Adults of all species that manage to raise young at times of krill shortage struggle to maintain the normal rate of food supply. Albatrosses must double the length of foraging trips and thus the interval between feeds for their chicks. Being flightless, penguins are unable to forage further afield and tend to return with smaller meals. Fur seals increase foraging time, diving deeper in pursuit of prey and resting for shorter periods. In the worst years, both chicks and pups may be abandoned, and any surviving offspring are left vulnerable and underweight. The Wandering Albatross *Diomedea ex-ulans* is the only top-order predator to breed on South Georgia that shows little variation in breeding success. Not surprisingly, it is the islands' only visitor that does not feed on krill, preferring an exclusive diet of carrion, squid and fish.

In recent years, scientists attempting to explain these spectacular krill fluctuations have tended to focus their studies far from the tiny land mass of South Georgia. It is well established that South Georgia's krill is not maintained by a self-sustaining local population but by the major krill breeding grounds of the Antarctic peninsula some 2,000 kilometres away. The Antarctic Circumpolar Current enables the krill to bridge this huge distance. Scientists working on the peninsula have uncovered substantial interannual variation in sea-ice cover. Every year the ice cover expands in winter, extending over as much as 20 million square kilometres around the continent, and contracting to just 4 million square kilometres in summer. The biological productivity of the sea-ice cover is exceptionally important. As the ice disintegrates each spring, the water is seeded by the vast algal communities that thrive at the interface between ice and water. These bloom as the retreating ice-edge is exposed to sunlight, creating the vast oceanic blooms that act as a focus for concentrations of zooplankton, sea-birds and mammals. Here under the pack ice, krill also find shelter over winter, feeding on algae while using the ice-ceiling as protection from diving predators. The winter sea-ice area fluctuates every year and with it the productivity which can vary by at least 50 per cent from year to year. Naturally, krill numbers tend to be highest when sea-ice cover is greatest.[10]

The mechanism by which El Niño affects sea-ice cover remains subject to debate. However, it is accepted that when the rain zone, normally situated in the western equatorial Pacific, moves east into the central Pacific during an El Niño, a series of atmospheric peaks and troughs emanate from the equator into both hemispheres. This creates a blocking high-pressure system to the north of the Bellingshausen Sea. Some scientists speculate that this weakens the normally strong southerly winds in the region, leading to a reduction in sea-ice to the west of the Antarctic Peninsula.

Flexibility is crucial to weathering El Niño's impacts at sea – and this applies as much to fishermen as it does to their quarry. Even the degree to which humans are affected can parallel the species they hunt. For example, fishermen with boats capable of pursuing migrating fish are those best placed to cope with the changes. Those worst affected tend to be tied to port or prevented from switching their catch through a lack of the necessary equipment.

In the days of subsistence fishing when communities had nowhere to turn but the seas, El Niños often signified a double dose of hardship, depriving fishermen of their catches and compounding the problem with appalling weather. For those living close to the margins, this unfortunate combination often spelled disaster. On some Pacific islands, the disappearance of fish went hand in hand with drought. 'It seemed as if the lagoon ecosystem was determined to combine with the terrestrial in ousting man from the scene,' wrote historian Henry Maude during the 1938 La Niña drought on Kiribati. 'At the height of the drought when the flora was dead or dying, the prolific fish population deserted the lagoon.'[11]

These days, hardship can be mitigated by management and a better understanding of the peaks and troughs that govern the availability and distribution of marine resources. Management agreements and treaties help balance the gains and losses that inevitably occur as a result of El Niño events. Treaties allowing access to the Pacific islands' fishing resource have been signed with large fishing nations such as Japan, Korea, Taiwan and the USA. One US treaty, for example, allows purse-seiners access to the fishing zones of sixteen Pacific countries. All member countries are guaranteed a small portion of the total fees paid, even if most of it hinges on the size of the catch, which can decline alarmingly as a result of El Niños.

The most commercially important fish are the various species of tuna. Nearly 70 per cent of the world's annual tuna harvest is hauled from the Pacific, the majority coming from the vast schools of skipjack *Katsuwonus pelamis*. Skipjack favour warmer waters. Although the warm pools themselves are relatively unproductive, enormous quantities of plankton aggregate at the oceanic margins of these warm pools, mainly as a result of being carried westwards with the prevailing currents.

The eastwards shift of the warm pool during an El Niño has several consequences for the tuna fishermen. The lowering of the thermocline, a temperature barrier that skipjack are reluctant to cross, pushes the fish below the reach of the seine nets. The composition of the catch can change too, with yellowfin tuna at times replacing skipjack. The costliest implication, both to the fishermen and treaty signatories, results from the tuna's tendency to migrate thousands of kilometres in pursuit of its preferred food supply. Although skipjack can be found throughout the tropical and sub-tropical Pacific, its best grounds in normal years are in the warm water pool west of 165°E. Once east of 170°W – roughly the longitude of American Samoa – catches are almost zero. In El Niños, the catch between 140°E and 160°E drops off considerably, with a concurrent increase in catch east of the Dateline. In the strong event of 1997, many boats fished east of the Line islands, situated around 160°W. In La Niñas, the trend is towards the reverse. For an island like Kiribati at a longitude of 175°E, normally on the eastern margin of the skipjack's range, the treaty proceeds during a La Niña may drop to almost nothing.

Elsewhere in the Pacific, it is a similar story of gains and losses. Bluefin tuna tend to shift their distribution away from Japan and extend all the way to the American coast. In strong El Niños, bluefin even venture into the south Pacific as far as Chile. Chile in fact benefits from the arrival of a number of species, particularly sardines deserting the Peruvian coast. In California the arrival of unusually warm waters is good news for recreational fishermen, with the appearance of tropical and sub-tropical visitors such as marlin, dorado, Pacific bonito and California barracuda. On the other hand, commercial fisheries suffer in a way that parallels the experiences of Peru's fishing fleet, with the normal catch of species such as whiting and salmon[12] both declining in number and moving to higher latitudes.

In parts of northern Australia's Gulf of Carpentaria, catches of banana prawn, *Penaeus merguiensis*, are positively correlated to years of high summer rainfall, a typical outcome of La Niña events. Intolerant of low salinity, adult prawns that breed in the Gulf's estuaries are forced, by the increased pulse of river runoff, to migrate offshore to within the sweep of trawlers. In the good years, the volume of catch has been so high that fishermen have struggled to offload the surplus. In the poor years, which tend to correlate with El Niños, export markets have been left unsatisfied.

For some species, El Niño's primary impact on fishermen comes not from its effect on the adult population, but from larval recruitment. The shrimp fishery off the Gulf of Mexico, one of the USA's most valuable fisheries, is affected by years of high river discharge, which in the northern part of the Gulf tends to occur in El Niño years. In this case the main effect is on the brown shrimp, *Penaeus aztecus*, and white shrimp, *P. setiferus*, fisheries. Life for the two species begins in the deep waters of the Gulf, where after three to five weeks, the larvae move inshore and mature into juveniles. After another four months, the prawns move offshore again. The size of the catch hinges on the survival rate of the larvae and juveniles, which is largely dictated by the atmospheric conditions. High summer discharge of the Mississippi River, characteristic of El Niño years, decreases larval survival, possibly as a result of a prolonged decrease in salinity. White shrimp catches during severe flooding years on the river, such as 1957–8, 1973 and 1993, tumbled by as much as 40 per cent.

During an El Niño, rock lobsters and several species of shrimp can appear along the coast of northern Peru as eggs and larvae, as well as adults, passively transported down the Ecuadorian coast by the unusual poleward currents. In the same region, the Peruvian scallop, *Argopecten purpuratus*, has been known to increase to very high densities – more than 100 large specimens per square metre.[13] As a consequence, several of the scallop predators such as octopus also proliferate. In 1982–3, the scallop harvest increased 4,000 per cent, spawning a highly profitable scallop export business in the Pisco area. The boom continued for three to four years until the scallops were fished out and other species managed to repopulate their former territory.

Over on the west coast of Australia, the degree to which the larvae of the western rock lobster *Panulirus cygnus* settle on to the coastal reefs, and the ultimate recruitment success of juveniles into adults, is closely correlated to El Niño years. The lobster supports the biggest single-species fishery in the country, with an average catch of over 11,000 tonnes, worth an estimated A$260 million. With variations of up to 20 per cent of the average in a bad year, the impact of an El Niño can be costly.

After settling on the reefs, the larvae develop into juveniles and grow for three to four years before migrating offshore as adults. Spawning occurs on the continental shelf's outer regions, and after hatching, the larvae float to the surface and are dispersed by offshore winds and currents across a large swathe of the south-eastern Indian Ocean. Several months later, the larvae begin their journey back towards the West Australian coast. At this point in their development, the larvae are still unable to swim effectively and they possess only sufficient energy reserves to survive a few more days. Somewhere over the edge of the continental shelf, the larvae metamorphose into the next stage of their development, enabling them to swim back on to the reefs. The trigger for this developmental stage is presently unknown, but is believed to be closely associated with the southward-flowing Leeuwin Current.[14]

The Leeuwin Current is very much a by-product of events in the western Pacific, and intimately linked with the same processes that drive the El Niño cycle. During normal years, the relatively high sea-level of the western tropical Pacific, associated with the build-up of a warm pool in the west by easterly trade winds, drives a flow of warm water through the Indonesian archipelago. This sets up the dynamic conditions for the Leeuwin Current, a relatively warm, low-salinity flow down the west Australian coast.

In El Niño years, when this warm pool dissipates across the breadth of the Pacific, the current is naturally diminished in strength, and sea-levels along the West Australian coast are lowered. Without the current's usual assistance, less larvae return to settle on the reefs, and the result is a dearth of recruits to the lobster fishery three to four years later, costing the industry up to A\$50 million.[15]

The repercussions of El Niño's disruptions to marine eco-systems may linger long after the El Niño itself has disappeared. Lasting changes may be detected in the age structure or distribution of a number of different species, or in the disrupted relationships between predator and prey. Young that experienced poor growth in their first few months may be hampered by reduced fitness for the rest of their lives. Populations may take several years, and in the case of exceptionally strong events even decades, to recover fully. Some effects, such as the estimated loss of 80–90 per cent of the 10–12 million sea-birds on the Pacific's Christmas Island during the 1982–3 El Niño, may take centuries to remedy. In the case of a handful of species, El Niño's impacts may not even become evident until long after the event.

It is October, two years after an El Niño. As the light fades over the Great Barrier Reef's Raine Island, the hulking, dark carapaces of green turtles, *Chelonia mydas*, first appear at the water's edge. Nervous at first of any unexpected movement, the turtles begin their arduous haul up the beach, their flippers rowing in unison as they drag themselves forwards. Every few minutes they pause to rest, their loud snorting betraying the strain of moving without the medium of water to support their weight.

To reach this spot, the turtles have travelled from feeding grounds dispersed throughout the Arafura and Coral Seas where they feed exclusively on sea-grass flats and algal-rich reefs. This year the numbers are exceptionally high, with around 12,000 filling the tiny island like humans filing into a sports stadium. Nesting space is at a premium, and some females dig where others have already laid their eggs, crushing the first clutch in the process. Their task completed, those with the strength remaining struggle back to the water's edge before dawn.

Many will repeat the ghostly procession up the beach several more times before the breeding season is over. In other years, this awesome spectacle becomes a paltry affair, with perhaps as few as two dozen turtles struggling ashore to breed. And the vast fluctuations in turtle-breeding numbers are not just confined to Raine Island. Right across the turtles' rookeries, as far afield as Java, these colossal variations are witnessed simultaneously.

Green turtle breeding is a long-drawn-out affair. Each year, only a portion of the adult female population enters the breeding process. In the cooler waters typical of the western tropical Pacific during an El Niño year, the turtles' vegetarian food thrives, allowing more females than normal to deposit the fat reserves that trigger egg production. Females with sufficient fat reserves take a further nine months to deposit several kilograms of yolk into the follicles that have been created for the breeding season. Then follows a long migration to the breeding grounds, courtship, ovulation and mating, extending the whole breeding preparation period to around two years. Although the physiological changes that males undergo in preparation for the breeding season are unknown, it is assumed they too are governed by a similar time-scale, because breeding males are also abundant two years after an El Niño event.

So, just like its land cousin, the tortoise, the green turtle seems to uphold the family reputation for tardiness by responding long after everyone else.

8

An Evolutionary Force
Imprints on the Natural World

'. . . as though by a miracle [rattlesnakes] suddenly appeared by the thousands in the fields because of the drought. Long, slithering, writhing, their heads triangular, they abandoned their lairs and they, too, migrated, like the human folk: and in their flight they killed children, calves, goats, and had no fear of entering settlements in broad daylight in search of food.'
Mario Vargas Llosa, on the 1877-8 drought in North-East Brazil[1]

All was not well with the waved albatross *Diomedea irrorata* of the Galápagos Islands. When biologist Catherine Rechten arrived at her study site in March 1983, she expected around 480 pairs to be returning from their feeding grounds off the coast of South America. All she could find was forty-two. Although Rechten didn't know it at the time, far out to sea the albatrosses' food source of fish and squid had virtually disappeared. But for the few birds strong enough to return for breeding, that was only the start of the crisis.

For over three months the islands had been lashed with rain. The bare strips of ground on which the birds normally landed were blanketed with a lush carpet of vegetation. Some only attempted a descent after more than a hundred trial approaches. There was heavy rain, but no wind, on which the birds rely for soaring and breaking their landing speed. Albatrosses find landing difficult at the best of times, and without the wind assistance, the birds were just crashing into the shrubs. Rechten found it painful to watch.

Each of the forty-two pairs produced an egg. But as the days rolled on, and the rain continued to bucket down, the damp ground turned into puddles and the puddles turned into lakes. Some parents were observed vainly trying to shovel their eggs to higher ground, others swimming hopelessly over the top of their submerged clutch. All males, except one, failed to return in time to relieve their incubating partners. Most eggs were abandoned, and eventually Rechten's study population was reduced to a solitary breeding adult. And it wasn't even incubating its own egg.

The first sign in the Galápagos Islands that something strange was afoot in the ocean probably occurred on 10 July 1982, when hundreds of thousands of crabs, *Euphylax dovii*, invaded the island of Genovesa. By the following day, they had all gone. The leading edge of the first Kelvin wave passed three months later, marking the beginning of several months of higher surface temperatures of 4–5°C above normal. Accompanying the warm water were heavy swells, rough seas and a sea-level rise of over 30 centimetres. By January, the normal upwelling of the cold equatorial undercurrent had been suppressed and was now 250 metres deep, as opposed to a norm of around 40 metres. This resulted in a dramatic hike in the temperature at 100 metres to 27°C – 12°C higher than normal and only 1°C cooler than at the surface. The whole archipelago was bathed in warm, clear blue water.

On land, the rain started in the middle of November, several weeks ahead of the normal wet season, and continued incessantly until July, three months after the rainy season usually ends. On the island of Santa Cruz, more than 3,225 millimetres fell over seven months, compared to an annual average of below 400 millimetres. In fact, more rain was recorded on the wettest day than normally falls in a whole year. The south-east trade winds disappeared, and the pleasant temperate weather normally enjoyed by the islands was replaced by typically tropical temperatures and humidity. For most terrestrial species used to semi-arid conditions, it was a time of extraordinary exuberance, with rates of growth, reproduction and population densities achieving records never before recorded on the isles.

The most obvious change to the marine life was the dramatic disappearance of several key species. By October 1982, the archipelago's waters were virtually devoid of sharks, turtles, groupers, whales and dolphins. Sea-birds such as swallow-tailed gulls, *Creagrus furcatus*, and blue-footed boobies, *Sula nebouxii*, had joined the exodus too. Most of the sea-birds that attempted breeding, such as great frigatebirds, *Fregata magnificens*, and masked boobies, *Sula dactylatra*, later abandoned their eggs and young, while the few chicks that survived exhibited very poor growth. Numbers of Galápagos penguins, *Spheniscus mendiculus*, dropped by 75 per cent and flightless cormorants, *Nannopterum harrisi*, by 50 per cent.

The breeding season of the Galápagos fur-seal *Arctocephalus galapagoensis* was a total failure. Fur-seal mothers, like all otariids – the group of eared seals that include fur-seals and sea-lions – need to intersperse their nursing bouts with periods of feeding themselves, and are consequently highly susceptible to a downturn in the food supply directly offshore. When their fish prey declines – as it did in late 1982 – mothers have only limited scope to change their behaviour to obtain sufficient food to keep their pups supplied with milk. In comparison, female phocids, or true seals, accumulate the energy for lactation during the months preceding the breeding season, and are much better insured against regional or short-term food declines. In 1982, Galápagos fur-seal mothers were forced to stay away on foraging trips for up to ten days, instead of the usual one to four. Even then, most were unable to meet the energetic requirements of lactation. Emaciated pups lined the rookeries bleating for their mothers, some attempting to steal milk from any passing teat. Most were abandoned. Not only were the four youngest year classes – or age groups – almost lost entirely but 30 per cent of the adult females as well. Less is known about the impact on the Galápagos sea-lion, *Zalophus californianus*, but in one monitored colony most of the pups had been abandoned by April 1983 and left to starve to death.

Further along some of the island shorelines, another of the Galápagos' endemics was locked in a similar struggle. Marine iguanas, *Amblyrhynchus cristatus*, are the world's only true marine lizards. These 'imps of darkness', as Darwin referred to them, specialise in feeding on the red and green algae that carpet the intertidal and subtidal zones. Large males can dive up to 15 metres to graze, whereas small females and young are restricted to the intertidal zone. In 1982, the change in water temperature spelt the disappearance of the normal red and green algae and their replacement by a species of brown algae, whose cellulose is indigestible to the microbes in an iguana's stomach. Coupled with the high sea-levels and heavy swells that restricted access to their feeding grounds, the iguanas starved. Within six months, only 30–40 per cent of the Galápagos population remained alive. All new hatchlings were exterminated, and it was not until late 1984 that there was sufficient food for females to accumulate reserves for egg production and to replenish the depleted population.

While those dependent on the seas for food were suffering, the situation for most land-based species was quite the reverse. Crops of seeds and fruits, as well as populations of most insects and spiders, were several times higher than normal, allowing species such as finches to breed prolifically. Mosquitoes, *Aedes sp.*, flourished in the salt ponds formed by the high tides, incessantly biting the relatively few fledgling seabirds that survived. Land iguanas, *Conolophus subcristatus*, devoured the abundant vegetation, resulting in high hatchling success and low mortality.

For some species the extra rain was not so welcome. Many mockingbirds, *Nesomimus parvulus*, on the island of Genovesa fell sick through what was probably pox, their legs swelling and feet and claws blistering. With many of the adults of this social species dying, juvenile survivors scattered across the island, terrorising finch parents and eating their nestlings in the process. Giant tortoises, *Geochelone elephantopus*, became extremely nervous of the sheets of water and raging streams that cascaded through their highland homes, many choosing the long and slow exodus down to the coastal plains where they fed on the abundant fruits and weedy vegetation. A few plants suffered too. Ancient *Opuntia* cactus trees toppled over, their roots saturated, their trunks engorged with water and overladen by the weight of smothering vines.

The world over, there are ecosystems like the terrestrial Galápagos for which an El Niño or La Niña event may be the cue for an extraordinary pulse of life. Some arid and semi-arid ecosystems may sit poised for years on end, full of plants and animals that are capable of breeding yet awaiting some particular non-seasonal cue before they can explode into frenetic activity. With El Niño's rains come strong bursts of plant production, seed banks are replenished, insects proliferate, and so on as a pulse of energy is passed up the food chain. Where the rains immediately follow drought, as is so often the case in the El Niño–La Niña cycle, numbers of predators may be depressed, allowing opportunistic species to breed quickly. The flipside of the coin, however, is that the opposite phase of the cycle may mean several months of deprivation, breeding failure and mortality. So, in effect, El Niño is the clock whose timing determines some of nature's rarest and most spectacular sights.

The coastal deserts of Peru and northern Chile explode into a carpet of flowers. Bumper yields of seed and herbage lead to outbreaks of rodents, or *ratadas* as they are known by locals in South America, followed months later by increased numbers of predators such as owls, hawks and foxes. Additional carrion, resulting from the change in conditions, may enable condors of the foothills and coast to breed where they have failed between El Niño years.[2] In East Africa, the saline lakes are filled by the heavy rains, attracting hundreds of thousands of lesser flamingoes, *Phoeniconaias minor*. The vast flocks feed on the algal soup, and when water levels have declined sufficiently to allow nesting breed on the soda mudflats. Rain falling on northern Mexico triggers tens and sometimes hundreds of millions of painted lady butterflies *Cynthia cardui* to migrate north across America into western Canada.[3] Drought forces rodents in the USA's Rocky Mountains to migrate to lower altitudes in search of food. Venomous western rattlesnakes follow their prey downslope, taking them into more populated areas, resulting in more human snakebite cases.

In the rain forests of Indonesia and Malaysia, the El Niño cycle appears to be the trigger, at least indirectly, for one of nature's true spectacles. Every few years, vast swathes of forest canopy simultaneously erupt into a patchwork of extraordinary colour. The phenomenon lasts for up to four months, with the flowering of each particular species taking up to three weeks, and is followed a few months later by mast fruiting. Synchronous flowering in itself is not unique, but where it does occur elsewhere, it is restricted in each case to a single species. In South-East Asia, the sheer scale, both in area and numbers of species involved, makes it exceptional. In the most prolific years, flowering can cover hundreds of square kilometres. The most spectacular in living memory occurred in 1983, stretching across the mountains and valleys of central Borneo.

The family of trees known as dipterocarps lies at the heart of the phenomenon. The family boasts some of the tallest and most beautiful trees of the South-East Asian rainforest, dominating the canopy and constituting 10 per cent of all tree species, and over 80 per cent of the emergents – those exceptional individuals that poke their crowns above all others. Most of the dipterocarps, along with up to 200 species of several unrelated tree families, are involved in the mass blooming that may feature over half the mature individuals and nearly 90 per cent of the canopy species. One tree is known to have opened 650,000 flowers in a single day, and with some dipterocarps able to produce up to a total of 4 million flowers, from which up to 120,000 fruits may be set, the forest is awash with billions of fruit.[4]

This saturation of fruit produced at irregular intervals has been suggested as one of the primary causes of the phenomenon. With numbers of predators kept low during the irregular intervals between mast fruitings, the sudden and unexpected surfeit of food quickly swamps and satiates seed predators, thereby enhancing the chances of seed survival.

Another theory is that pollination is higher because the mass flowering results from the exponential increase in the immigration and breeding of insect pollinators, while at the same time the slightly staggered flowering between species avoids pollination from foreign pollen.

For a long time, the trigger for the phenomenon eluded scientists. But from recent studies, mass flowering events now appear to be set off by a drop of roughly 2°C or more in the minimum nighttime temperature over a few nights in a row.

In this part of the world, the occurrence of such climatic fluctuations is very irregular. The central part of the South-East Asian tropics is characterised as being completely aseasonal, lacking in an annual dry season, with the summer monsoon from the Indian Ocean followed by the winter monsoon from the Pacific and South China Sea. In such an environment, the most extreme conditions experienced are often associated with the El Niño cycle. But how the El Niño cycle causes the extended period in minimum temperatures is still not clear. On the eastern slopes of Sumatra, Borneo, lower nighttime temperatures may be associated with the dry air masses that descend on the region during El Niños. In other parts, such as the eastern and southern Malaysian peninsula, the correlation is rather with La Niña. One study showed that the climatic anomaly during La Niñas was actually over the Khorat Plateau of South-East Asia, which in turn forced short bursts of dry north-easterlies to be blown much further south than usual.

To the Dayaks of the Borneo interior, mast fruiting means much more than just a heavy bounty of food. The realm of the phenomenon is thought by some tribes to encompass the entire cosmos. Some even believe that the destruction of the current world and the genesis of the next will happen during the greatest ever mast fruiting. To the Dayaks, this exceptional natural bounty from the mast fruiting means the normal rules of fruit harvesting no longer apply. Collection switches from trees close to the villages to those deep in the forest thought to belong to the spirits. The fruit is gathered from the ground because these sacred trees cannot be climbed, unlike the individually owned village trees. As a result, not only is the whole community permitted to participate, but the proceeds are used to pay for special one-off purchases, such as heirlooms or school fees.

Fruit harvesting is not the only attraction for the Dayaks

during these periods. The bearded pig, *Sus barbatus*, the native pig of Kalimantan and the prized delicacy of the forest people, is known to congregate in enormous numbers during these times. When the exceptional masting of 1983 occurred, the massing of pigs at the headwaters of the Baram River was estimated to total at least a million. In another congregation in 1954, so many pigs were killed as they migrated across the Kayan River that thousands of rotting carcasses drifted downstream, coming to rest where the river widened at the town of Tanjungselor. Being Muslim, the local townsfolk were outraged, and war was declared on the Dayak hunters. It took the imposition of the Indonesian police and a small armed expedition to end the pig slaughter and pacify the residents.

Local hunters have observed that mass flowering is the cue for pigs to mate *en masse*, so that after a gestation period of three to four months, the piglets are born into a time of optimum food supply. In fact, local farmers maintain that if the sound of mating pigs is heard from their longhouses, they can expect a bumper crop of fruit in a few months' time.

According to Dayak beliefs, the forest spirits exact a price on humans for this natural bounty. These spirits, associated with the giant fruit trees in particular, are believed to see humans that enter the forest in the same way that humans regard wild pigs. By concentrating themselves in the forests during mast-fruiting periods, humans act like game in the hunting grounds of the spirits. Hunters and fruit gatherers are therefore thought to be much more vulnerable during this time, and according to at least one tribe, must pay the toll of 'one basket of human eyes'.[5]

An even greater massing and migration of animals known as *trekbokken*,[6] the Afrikaner term for the great movements of springbok, *Antidorcas marsupialis*, were recorded in southern Africa. Today numbers of this small antelope are a mere fraction of what they once were, and the spectacular *trek-bokken* no longer occur. But every ten to forty years until as recently as the twentieth century, huge numbers of springbok would mass together and attempt to reach better pastures. The historical accounts that describe this phenomenon conjure up extraordinary scenes.

Even allowing for exaggeration in the eyewitness accounts, the sheer number of animals involved must have been awesome. One observer near the Orange River in 1896 'computed the number to be not less than five hundred thousand . . . in sight at one moment . . . They extended twenty-three hours in one direction and from two to three in the other – that is, the whole trek occupied a space of country one hundred and thirty-eight by fifteen miles.'[7] Another account likened the sea of antelope to 'the flood of some great river'. One witness, describing the phenomenon in the same region in 1888, took a slightly more mathematical approach: '. . . About ten thousand can stand on an acre, and I can see in front of me ten thousand acres covered in buck. That means at least one hundred million buck. Then what about the miles upon miles around on all sides as far as the eye can reach covered with them?'[8] To enhance the spectacle still more, a roll call of predators and scavengers invariably trailed in the wake of the springbok. Lions, leopards, hyaenas, jackals and circling vultures kept the antelope in their sights to prey on those too weak to keep up.

The common denominator that appears to have precipitated each trek was drought, and in virtually all cases, the springbok were marching towards a region of higher rainfall. Animals were often noted for their emaciated and desperate state, throwing off their usual caution as they plunged recklessly into rivers and soldiered on through towns and across railway lines, devouring everything in their path. One Transvaal pioneer described how his family was rescued by his bushman guide from being crushed in a springbok stampede in 1891. At the first sign of a gathering dust cloud on the horizon, the guide advised the family to leave the river valley where they were travelling and take refuge in their wagon at the top of a hill. Even in this place of relative safety away from the main swarm of antelopes, piles of buck still crashed into the wagon, jamming the wheels. All around the family, antelope corpses mounted into a huge stack until the *trek-bokken* had passed. The pioneer later wrote that the pile of bodies constituted 'more biltong than he could have secured in a year's expensive shooting'.[9] Another trek in 1892, witnessed in Namaqualand inland of the south-west coast, ended in the Atlantic Ocean, where the buck 'dashed into the waves, drank the salt water and died. Their bodies lay in one continuous pile along the shore for over 30 miles, and the stench drove the trekboers who were camped near the coast far inland.'[10]

Throughout much of the summer-rainfall region of southern Africa, drought is strongly correlated to El Niño years. It is therefore not surprising that the recorded *trekbokken* years in the Karoo and Bushmanland often coincide with El Niño events. In the winter-rainfall zone of the south-west coast, where El Niño years tend to bring heavier rain, it is no coincidence that the few recorded treks of this region have tended to overlap with the dry La Niña years.

Inevitably the spread of Boer pioneers across southern Africa spelt the end for the *trekbokken*. Driven by desperation and seemingly oblivious to danger, the springbok exposed themselves to huge-scale slaughter. News of the *trekbokken* attracted excited hunters from miles around. When the magistrate of Prieska heard of one trek passing through his town, he could not resist abandoning his duties to sit down on the steps of the courthouse and shoot 'a few nice ones as they strolled past'.[11] Game management fences and the rinderpest epidemic of the 1890s also took their toll on the *trekbokken* and by the 1990s the migrations were reduced to just a few thousand animals.

When Charles Darwin collected his samples of finches from the Galápagos Islands, he could not have appreciated the light they would later shed on his revolutionary ideas on evolution and natural selection. He did not find it easy telling the different finches apart, or even recognising that some were finches at all. In fact he described them as an 'inexplicable confusion', bagging his samples together without bothering to label each with their provenance. It was only once back in London, with the help of the ornithologist John Gould, that Darwin began to realise that his specimens were a closely related group of birds that had evolved particular adaptations in response to different environmental niches provided by the many islands.

But while Darwin had identified the process of natural selection, he believed it to occur so slowly and imperceptibly that it could not be witnessed by humans in the course of real time. 'We see nothing of these slow changes in progress, until the hand of time has marked the long lapse of ages,'[12] he wrote. However, nearly a century and a half later, these same changes have been recognised as neither rare nor slow. Now another ground-breaking study of the Galápagos finches has recorded evolution in action, providing at the same time some powerful insights into the evolutionary implications of El Niños and other catastrophic events.

There are two main species of finch on the island of Daphne Major – the medium ground finch, *Geospiza fortis*, and the cactus finch, *G. scandens*. By 1983, a study run by Peter and Rosemary Grant since 1977 had virtually all finches on this tiny island tagged and measured. When the heavy rains of 1983 hit, there was an explosion of seeds and insects, particularly caterpillars which are favoured by young finches. And whereas in 1981–2 all sixty of the finch nests started were abandoned through a lack of food, in 1983 there were over 1,100 nests producing nearly 1,600 successful fledgings. In fact the finches bred several times over the breeding season. The most successful medium ground finch fledged twenty-five young, with the most successful cactus finch only slightly less prolific with nineteen. Finches were even observed pairing and mating as young as three months old, as opposed to the normal two years. Even with the relatively high hatchling mortality, resulting from the abandonment of many of the hundreds of nests started, nearly four times as many young were raised than in normal years. For many individuals, this one season constituted a substantial portion of a lifetime's reproduction.

Importantly for the finches, the amount of seed had not only increased by around eleven times on the previous year, but the composition of the seed bounty had also changed with smaller, softer seeds increasing from 20 per cent to over 80 per cent of the total seed biomass. This was partly the result of conditions being too wet for the large cactus trees, and the partial failure of their crop of large, hard seeds. When the overall seed supply began diminishing after the rains finally stopped in July 1983, the larger, harder seeds were depleted first. Imperceptibly to the naked eye, this change in seed size favoured the smaller medium ground finch individuals that were more adept at eating smaller seeds, and disadvantaged the bigger birds with their larger appetites. A full two years after the rains ended, mortality was still highest among the bigger individuals with bigger beaks. The shift towards smaller beak size, known to be a highly inheritable trait, was shown to have been transmitted into the subsequent generation. The beaks of these new recruits were found to be 0.12 millimetres narrower on average than those of the previous generation. This differential in beak size may seem minute, but variations of a fraction of a millimetre were shown to determine the difference between life and death. The accuracy of the measurements taken also enabled evolutionary processes to be captured in the course of a single breeding season.

The evolutionary shift to smaller body size was not the first time the study had detected an impact on selection as a result of climatic anomaly. During the earlier drought years of 1977–8 and 1980–1, the smaller, softer seeds had been the first to disappear, leaving a greater percentage of the larger, harder seeds. The finches with the larger bills that could crack them open were at an advantage, and the smaller individuals that could not were the most vulnerable. After the droughts, the medium ground finches that had survived were found to be 5–6 per cent larger than those that had died. Among the new generation, the average beak size was 4–5 per cent deeper than the average before the drought. In effect, the shift in direction of natural selection by the favourable rains of 1982–3 was only reversing an earlier shift in the opposite direction forced by previous droughts. What is more, this oscillation in natural selection was shown to occur even within the lifetime of an individual.

The fluctuations in food availability did not just affect natural selection, they also influenced sexual selection. The usual ratio of sexes in finches is roughly equal, but males tend to be much larger. Following times of food shortages caused by drought, more males survive than females because of their greater size, leaving the surviving females with a wider selection of potential mates. In such times, large males are heavily favoured, and as a consequence, tend to father many chicks. Small males tend to remain bachelors. Conversely, after heavy rains when bigger birds are at a disadvantage, more of the larger males die, leaving the smaller male survivors to mate with the females.

Selection of smaller individuals during climatically abnormal years in the Galápagos Islands has not been confined to finches. Finches occasionally pluck ticks from the backs of marine iguanas that also benefit from being smaller in certain conditions. Because smaller individuals are able to get their mouths closer to the rocks to crop algae, they can extract more food per bite than larger counterparts. In addition, possessing a relatively larger surface area, smaller iguanas warm up faster as they bask in the sunshine between foraging bouts, and can therefore extract more food per trip. During strong El Niño events when the supply of digestible algae declines dramatically, the bigger iguanas with lesser feeding efficiency are the first to starve. At the same time, very small individuals also die in disproportionately high numbers because they are unable to withstand the stronger wave action typical of El Niños that washes many off the rocks into the sea.

With males about twice the size of females, marine iguanas are strongly sexually dimorphic. In normal years, there is a marked sexual selection for larger males, allowing a relative handful of males to become very big. Larger males control the best territories, usually above the high-tide mark and closest to suitable nesting grounds, enabling them to achieve the most matings. In a typical group of twenty-five territorial males, for example, three or four males would normally perform 80 per cent of the copulations. Yet the same trait for gaining maximum access to territory and females also predisposes the large males to mortality whenever conditions deteriorate. In the same theoretical grouping of twenty-five males following an El Niño event, nearly all the matings might be performed by about 80 per cent of the males as opposed to a few individuals. In effect, natural selection away from larger body size during El Niño related food shortages acts to reverse the direction of sexual selection in more favourable years.

El Niños can have a similar effect on sexual selection for

another inhabitant of the rocky shoreline, the Galápagos fur-seal. Fur-seals are nearly as dimorphic as iguanas, with territorial males weighing around 65–70 kilograms, more than twice that of the average adult female. In order to maintain their shoreline territory during the breeding season, males must fight to retain access to females, enduring two to six weeks of fasting on land at the cost of significant weight loss. When they returned to the ocean to replenish their energy reserves during the 1982–3 breeding season, the normal food supply was not available. As a consequence, all territorial males starved to death. On one breeding site, five small, sub-dominant males were able to establish dominance over the whole shoreline, each basking in the control of a territory over four times larger than normal. Without the usual territorial vigilance, even diminutive males no bigger than the females were able to sneak multiple copulations.

In 1982–3, while smaller iguanas, finches and fur-seals were being favoured by El Niño conditions in the Galápagos, further north another shift in sexual selection was also being witnessed. During the summer months on the islands of the Bering Sea, the red fox, *Vulpes vulpes*, normally preys on the sea-birds that return to nest on the islands' vertical cliffs. Carefully negotiating the jagged rocks and slippery ledges, the foxes scamper up and down the cliffs, hunting both adults and chicks. Eggs are stolen too, cached away to augment the meagre supplies on offer during the harsh winters.

In 1982, the numbers and composition of sea-birds that returned to Round Island were drastically altered. Very few black-legged kittiwakes, *Rissa tridactyla*, and common murres, *Uria aalge*, the most common prey of the fox because they nest in colonies on exposed ledges, came back.[13] Instead, the foxes were forced to switch to the smaller prey of parakeet auklets, as well as a few horned and tufted puffins. These three species were also reduced in number and their scattered nests, hidden in rocky crevasses and among the scree slopes, were harder to find than the foxes' normal prey.

Relative to many other red-fox populations, the foxes on the Bering Strait islands normally enjoy an abundant summer food supply with the density of their prey linked to the rich productivity of the offshore marine ecosystem. Accordingly, the concentration of foxes on Round Island is far higher than on the mainland. This richness of food supply also allows most fox groups to adopt a polygynous mating system, with more than one female sharing the same male. Although uncommon in birds, polygyny occurs wherever the benefits to a female of sharing a male of superior fitness or territory outweigh those of establishing an exclusive bond with a single, inferior, unmated male. With the scarcity of nesting birds in 1982, all mating became monogamous, implying that the task of pup-rearing was too onerous for a mother without the resources of a monogamous partner.

Another mechanism by which El Niños may influence evolution lies in their impact on species' distribution. Some populations are forced by the change in conditions to shift beyond their normal range, merging with other normally separate populations. Others may settle into new territory, leading to a splintering into disjunct sub-populations, and in the rare cases where prolonged isolation results, even speciation. In exceptional cases, where mass mortality during an El Niño occurs in conjunction with other stresses on the population, extinction may become a serious threat.

Many stories have emerged from recent El Niños of changing currents and patterns in food availability extending the ranges of various species. For example, bottlenose dolphins, *Tursiops gilli*, normally found off southern California, appeared in San Francisco Bay in 1983. Likewise, triggerfish, *Melichthys niger*, were found in Alaska, 2,800 kilometres north of their former northern record. In northern Chile, a group of South American fur-seals arrived in 1982 having migrated from Peruvian waters where fish stocks had collapsed, and subsequently settled and bred.

The dispersal of invertebrate larvae is also known to be aided greatly by the anomalous currents of an El Niño. Each species has a set period of time in which it is competent to settle and metamorphose. By increasing the speed of certain currents, larvae are therefore able to spread further. Colonies of *Acropora spp.* coral are found in a few sites around the eastern Pacific, where they are presumed to have travelled from ancestral reefs in the central Pacific. As the competency period of *Acropora spp.* coral is less than twenty-five days, and the nearest reefs would normally take at least fifty days to reach via the North Equatorial Countercurrent, it has been suggested that these journeys were completed during exceptional El Niño conditions.

Anomalous winds and heavy storms can have a similar effect, blowing birds and insects off course. Heavy floods may wash animals and plants out to sea on floating debris. This may have been the mechanism by which two carabid beetles, *Cicindela trifasciata* and *Pentagonica flavipes*, were transported from the west coast of South America to the Galápagos Islands, where they first appeared in 1983. It is highly probable that giant tortoises were first transported to the Galápagos in a similar fashion, flushed down a raging river on the Ecuadorian mainland and out to sea on a raft of vegetation.

In the vast majority of cases, individuals that are blown, carried or attracted beyond their normal range either die from the competition for resources in their new territory or return to their normal range when conditions improve. In some cases though, the peculiar climatic conditions of an El Niño enable the emigrants to gain enough of a foothold in their new environment to consolidate their position. During 1982–3, several large ground finches, *Geospiza magnirostris*, landed on Daphne Major. The annual appearance of a few stray individuals on the island is not unusual but the finches had never succeeded in breeding. However, under the exceptionally favourable conditions that year, the large ground finches were able to breed, and their offspring survive on Daphne Major to this day.

The smooth-billed ani, *Crotophaga ani*, remained rare on the Galápagos for two decades after being introduced in the 1960s from mainland Ecuador. The bird, brought to the islands by farmers in the mistaken belief that it might relieve the tick problem in cattle, has extended its range significantly during recent El Niños, and now numbers some 1,500 individuals. Similarly, the hot, humid weather of an El Niño year in the Galápagos Islands provides perfect nesting and foraging conditions for the fire ant, *Wasmannia auropunctata*. This vicious, biting insect, regarded by humans as a serious nuisance, has been able to make several incremental gains in its range during El Niños after being accidentally introduced to the archipelago.

In the dry, coastal deserts of Peru and northern Chile, small islands of vegetation known as *lomas* formations bring relief to an otherwise colourless landscape. These green oases, each unique in their species composition, survive the desert conditions by subsisting on the winter mist. El Niño events bring the only rains to fall in these deserts, stimulating an explosion of flowers, and attracting livestock and guanaco herds down from the foothills. The infrequent rains present bridging opportunities between the isolated communities, creating the scope for them to expand and merge with others, before the return to harsh, arid conditions. The simultaneous flowering of all the *lomas* plants during El Niño rainfall also creates potential for genetic reshuffling between communities.

In the Ecuadorian rain forest, a study involving nearly

4 million arthropod specimens has shown that the dry conditions that follow El Niño years may have important implications for maintaining genetic variability and high species diversity in insects and spiders. Without the limiting effect of rainfall on egg hatching and the time adults can spend on food plants, arthropod populations expand in number and range, creating the opportunity to establish themselves in new territories before unfavourable conditions return.[14]

For many other faunal groups, El Niños have the reverse effect, leading to shrinkage in range and, in exceptionally severe cases, the loss of genetic variability through a bottleneck effect. However, under normal circumstances, total extinction is an extremely unlikely outcome and most populations bounce back when favourable conditions return. In spite of this, there is recent evidence to suggest that when adverse El Niño conditions are superimposed on to existing stresses, a natural contraction in distribution may expose a species to a real risk of extinction.[15]

Climatic fluctuations associated with recent El Niños have been attributed to the mysterious decline of a number of species on at least two continents. Amphibians, with their moist permeable skins and tendency to lose water rapidly in raised temperatures, are particularly sensitive to drought. In the Monteverde Cloud Forest Reserve in Costa Rica, twenty species of frogs and toads disappeared following the 1986–7 El Niño. The golden toad, *Bufo periglenes*, renowned for its striking orange day-glo colour, was endemic to the reserve, but has never been seen since. The 1986–7 El Niño brought record dry, warm conditions to Monteverde, compounding a trend first noted in the mid–1970s towards climate change on tropical mountains. This warming, associated with the gradual rise in global sea-surface temperatures, is thought to have raised the level of cloud formation in the cloud forests of Costa Rica. The resulting loss of moisture, combined with the dry conditions imposed by El Niño, appear to have signalled the final days for the golden toad.

Several frog species in Australia have suffered similar fates. The once locally common southern corroboree, *Pseudophryne corroboree*, found mainly in undisturbed habitat in the Snowy Mountains, is now thought to be on the edge of extinction. The tadpoles of this strikingly colourful species are particularly susceptible to drought because of their need to over-winter in water. In summer, the adults seek out breeding sites in shallow sphagnum bogs and pools that become flooded after winter rains. Dry conditions can empty the pools and kill the developing tadpoles. Incremental declines in corroboree numbers have been recorded following a series of droughts, particularly after the strong 1982–3 El Niño event. All low-altitude populations have now disappeared, and the species clings to a few wet sites.

It is highly unlikely that El Niño related droughts are the sole cause of these amphibian declines. However, when added to a cocktail of other potential stresses such as parasitism, pollution, ultra-violet radiation through ozone depletion, or global warming, these climatic extremes can prove particularly deadly. One possible extra stress that has generated much interest among scientists is the role of a *chytrid* fungus that has been found on dead individuals of several frog and toad species in Australia and central America. It is suspected that host susceptibility to this fungus may be related to changing environmental conditions.

Studies of population changes in a number of species during recent El Niños have shown how severe natural disturbances can cause micro-evolutionary change in the course of a single event, such as witnessed with Galápagos finches. Over a much longer time-scale, the intense selection for certain traits that enable a species to survive El Niño extremes may emerge as a major evolutionary force. The question that must first be answered is whether El Niño has been around long enough to shape such adaptations.

Like a mature actress, El Niño has been wary of revealing its age. Some scientists have suggested that it only evolved into its present form around 7,000 years ago. This date is based on evidence from shell middens from the northern Peruvian coast that have revealed the presence of tropical molluscs and fish before 5000 BC. Today only cold-adapted species would be expected in these waters. Some evidence from Australia and elsewhere appears to corroborate that the Pacific Ocean underwent some major change in oceanographic conditions around this time, lends weight to the theory of a 5000 BC El Niño start-date. Other scientists have dismissed this argument, believing that tropical and temperate species could have co-existed at the time. Furthermore, sedimentary and soil samples that date back more than 40,000 years from the same region suggest that catastrophic flooding, indicative of strong El Niño events in Peru, has been occurring for at least as long. But these findings may have been overshadowed by a more recent discovery of fossil coral from Sulawesi. Analysis of the coral has shown that interannual fluctuations, very similar to those observed in modern instrumental records, have been occurring in the Pacific for at least 125,000 years, right through the last interglacial period.

There is a wealth of circumstantial evidence to show that within regions of strongest El Niño impact, the biota displays marked adaptations to climatic varability, strongly implying that El Niño has been operating for a very long time. In fact, adaptations to environmental unpredictability among species and communities within these areas may provide some of the best evidence of El Niño's longevity.

Many of the life-history characteristics of tropical pelagic sea-birds – considered extreme relative to other birds – are now thought to be adaptations to periodic breeding failure.[16] Many live thirty to fifty years, enabling adults to experience several failed breeding years in their lives while still achieving reproductive success. Clutch sizes are small, with only one or two eggs. Chicks grow very slowly, staying dependent on their parents for longer, and maybe not even breeding until at least five years old. Fledged juveniles of some species move far away from their birth-site and the feeding range of the adults. In a year of high adult mortality, the sub-adults may find better feeding conditions than their parents, and may be able to repopulate the decimated colonies upon their return. When the differences between seabird life history characteristics and those of land birds were first recognised, they were presumed to be determined by the difficulty seabirds have in finding food from the open ocean and transporting it long distances back to feed their young.[17] But more recent studies have shown that the birds are not operating at the edge of their energetic capacity. This reserve enables them to alter their feeding patterns if conditions deteriorate, allowing an increase in feeding effort while continuing to feed their chicks even when their prey becomes harder to find.

One of the physiological traits that a marine iguana employs to increase his chances of surviving a downturn in algal supplies may be another example of El Niño adaptation. Because it is an advantage to be smaller when food is scarce, iguanas have evolved the ability to shrink. When the algae become abundant again during La Niñas, the iguanas are able to assume their normal body length. The only other documented case of shrinkage of bone-length in vertebrates belongs to astronauts subjected to prolonged periods in gravityless conditions in space. But in these instances, the contraction in bone-length has not been reversible.

Some degree of shrinkage in iguanas may be explained by a reduction in cartilage tissue. However, because the amount of shrinkage can be as much as 20 per cent of body length, it is believed that most of it results from bone absorption. Larger individuals have been shown to shrink more, and those individuals that shrink the most have been found to survive longer. As very strong El Niños tend to kill the larger iguanas before they get a chance to reduce their size, shrinkage is thought to be an adaptation to the more frequent, moderate El Niños.

Possibly the most interesting insights into suspected evolutionary responses to El Niño are found in Australia. Throughout most of this continent, the overwhelming influence on rainfall stems not from seasonal fluctuations but from El Niño. In a land where drought and flood are ever recurring themes, much of the fauna and flora exhibits distinctly characteristic adaptations to highly variable and unpredictable rainfall.

Australia's avian fauna, for instance, display characteristics that are significantly more prevalent than in birds anywhere else. Over a quarter of Australia's birds are nomadic, enabling them to track the most favourable conditions. As a consequence, concentrations of these species, such as the emu, vary considerably throughout the interior from year to year. Around 12 per cent of the continent's breeding residents and regular migrants have been recorded breeding co-operatively – whereby individuals forgo breeding themselves and assist others in the rearing of their young – a proportion five times greater than the world average. In an unpredictable environment, the high costs of reproduction during an unfavourable breeding season restrict an individual's chances of successful breeding, while increasing the relative benefits of remaining a non-breeder and helping within its natal group. Australian birds also demonstrate huge flexibility in breeding behaviour. The timing of breeding, together with the number of eggs in a clutch or broods produced in succession, can all vary enormously. Even the age at which birds breed is subject to change. These traits allow birds to breed rapidly and exploit conditions when they are good, and to conserve energy when they deteriorate.

The zebra finch, *Taeniopygia guttata*, boasts many of these flexible features. It usually forms flocks of around 50–100 birds, but will congregate in thousands in dry conditions. In the inland areas of its range, breeding is completely aseasonal and totally dependent on the unpredictable rainfall. Within hours of the arrival of rain, the finches begin courting and nesting. And within just nine to ten weeks, the young are capable of breeding. If conditions remain favourable, the birds have been known to breed continuously for eleven consecutive months.

The banded stilt, *Cladorhynchus leucocephalus*, is an equally opportunistic breeder after bouts of rain. In order to breed, the stilt requires freshly filled salt lakes dotted with islands. These are found in inland areas but are only flushed with water by the rare but prolonged downpours generally associated with La Niña years. Prior to 1989, the stilt's breeding behaviour had seldom been witnessed. In the weeks that followed the drenching rains in the Flinders Ranges in early summer, waters streamed inland down the rivers and creeks, filling the normally empty Lake Torrens. Before the first waters had even trickled into the lake, stilts were witnessed arriving from virtually all over the country.[18] A host of other species, including waders, waterfowl,[19] raptors and scavengers, also flocked to the shoreline, snatching the opportunity to feast and breed.

Billions of tiny insect, mollusc and crustacean eggs produced during the lake's last replenishment many years earlier lie dormant in the dry salt beds. Flushed into life by the arrival of the water, the larvae are joined by countless immigrants, such as fish from nearby billabongs and muddy pools, transforming the lake into an aquatic smorgasbord. The most essential component of this productive soup is the Australian brine-shrimp, *Parartemia spp.*, a salt-tolerant crustacean on which stilt chicks are totally dependent. In 1989, up to 100,000 stilts appeared at Lake Torrens, wasting little time in scraping a shallow depression in the ground into which to lay their eggs. The full breeding cycle is as short as seven weeks, with chicks capable of flying only three weeks after hatching. But in order to maximise this ephemeral opportunity, chicks are pushed from the nest before fledging to join thousands of others in a floating crèche supervised by male escorts. At some point as the water evaporates, the salinity of the lake increases too much even for the salt-tolerant brine-shrimp, and the stilts' banquet is over. These long-lived birds, together with any chicks that manage to fledge before the food supply disappears, fly off towards the coast to wait several years before the next mysterious cue signals that rains have again filled their inland lakes.

In contrast, the opportunistic breeding behaviour of the long-haired rat, *Rattus villosissimus*, has been recognised since the nineteenth century. It was first noted by the explorers Burke and Wills, on their fateful trek through Australia's interior. Upon arriving at Cooper's Creek for the first time, the expedition was immediately overrun by a plague of rats. According to Wills's meticulous diary, the rats attacked their stores 'in such numbers that we could save nothing from them unless by suspending it in the trees'.[20] Wills may even have betrayed the indirect cause of the plague by observing that a flood had recently inundated the region. 'The logs and bushes high upon the forks of the trees tell of the destructive floods to which this part of the country has been subjected, and at no very distant period.' Small numbers of rats manage to survive long droughts by finding refuge around the few permanent waterholes. With the flush of new growth that accompanies a flood or heavy rain the rat is able to irrupt into the surrounding habitat. Plague numbers can be achieved rapidly, with gestation taking just three weeks, an exceptionally short period for a placental mammal. The young are sexually mature after only seventy days. After a time lag, the rats' predators – raptors, dingo, snakes, foxes and feral cats – breed in great numbers too, eventually helping to signal the declining phase of the rat's population cycle.

One predator in particular, the letter-winged kite, *Elanus scriptus*, is so dependent on irruptions of the long-haired rat that its breeding is tied in phase with it. Normally the kite is a solitary bird, but while rats are abundant, flocks gather, and any anti-social tendencies are suppressed. Nesting space quickly reaches a premium, sometimes with more than one female using the same nest simultaneously. Not only do the rats provide the birds with regular meals, but the fur of their carcasses is used to line the nests.

When drought strikes, there are two main strategies that animals and plants can employ to cope. The first, used by the long-haired rat, for example, can be characterised as drought evasion. Birds achieve it through their mobility, flying away from a drought-affected area, sometimes as far away as the coastal fringes of the continent. Invertebrates produce eggs and larvae that may lie dormant for many years. Burrowing frogs bury themselves deep into the soil, sealing themselves against desiccation with a skin-cocoon. Ephemeral plants that germinate after rain set their seed and die. Shielded from the ravages of the drought wherever they may be, drought evaders more or less sit it out and wait for better times.

The second strategy is best described as drought tolerance. Social ants and termites – highly successful groups in the Australian environment – cease external activities and live on stored grain in their communal galleries. Long-lived perennial plants like mulga are hardy enough to withstand all but the severest drought, waiting for the rare bouts of prolonged rains that trigger flowering and seeding, and later enable their seedlings to establish. Other species rely on their mobility to find transient patches of food that have sprung up in response to localised rainfall. Small marsupials and rodents can travel relatively vast distances, several times further than similar-sized mammals on other continents.

When rains signal the end of a drought, the red kangaroo has made a head start in breeding. Already, the female is carrying a fertilised egg waiting for its climatic cue. Several months beforehand, a fertilised egg would have developed into a bundle of cells – a blastocyst – that would have lain dormant inside the mother until it received its hormonal wake-up call. Once triggered by the expulsion of the oldest joey from the pouch, the new joey can be on the teat within thirty days. And by the time that joey is bounding around outside its mother there will be a younger one inside the pouch, and another egg awaiting its cue. Conception to independence takes around 600 days but by having up to three young on the reproductive conveyor belt at the same time, females can produce one independent young every 240 days, making the most of the good times while they last.

But this strategy is often cruel to the young. Breeding is continual, and unless there are favourable rains, mothers are not able to supply the milk necessary for all their offspring to survive. The first teat to be switched off is the fat-rich milk needed to sustain the mobile joey, and as a consequence, many young die in transition from the pouch to independence. For a while mating continues, and new young are born into the pouch. But once drought worsens, the mother scales back her investment in breeding even further. The carbohydrate-rich milk necessary for development of the pouch-bound joey is the next to dry up. Finally the matings cease and the mother waits for the next rains and the flush of new growth.

With its quiescent blastocyst providing a head start in the race to take advantage of improved conditions, the red kangaroo takes opportunistic breeding to an extreme. But the critical point about marsupial reproduction, as opposed to that of placental mammals, is that mothers make a very small energy commitment to the embryo, with most marsupial embryos weighing less than a gram at birth. As a consequence, the marsupial can cut her losses with minimum loss of investment. In an unpredictable environment, where unexpected losses of young must occur frequently, the facility to terminate parental care without great cost to the parent is a distinct advantage.

Until recently, it was commonly believed that marsupials were only able to persist and diversify in Australia because of the continent's long isolation from competition from placental mammals. The more 'primitive' marsupials, or so the theory went, would be out-competed by the inherently superior placental mammals, a view no doubt enhanced by our own 'placentocentric' perspective. The fact that relatively few marsupials on the South American continent survived the aftermath of the union with the placental mammals of the North American landmass, when both continents joined some 2.5 million years ago, was taken as evidence that marsupials would not survive in any significant numbers alongside placental mammals.

But in recent years there has been a general shift away from this view. A reassessment of the paleogeographic evidence from South America has shown that roughly the same proportion of that continent's indigenous placental mammals went extinct as did South American marsupials. In addition, the recent discovery in south-eastern Australia of a fossil of a 120-million-year-old placental mammal implies placentals became extinct in Australia while cohabiting with marsupials. It is now presumed that certain traits of marsupials have enabled them to survive better the peculiar characteristics of the Australian environment, particularly its distinctly poor soils and unpredictable climate. In such an environment, where rainfall periodically declines and resources become severely limited, a premium is placed on those with low energy requirements.

It has been thought for some time that marsupials have a relatively lower metabolic rate than placentals – on average 30–40 per cent lower. Previous explanations to account for this have focused on either their perceived 'primitiveness' or the generally low nutritional quality of food available in most Australian habitats. One recent study on metabolic rates of mammals has suggested that zoogeographical reasons may have been a prime determinant. Marsupials, the study confirmed, do have lower metabolic rates, but so do placentals from the Afrotropical and Indomalayan regions – both regions also strongly influenced by the El Niño cycle. In these regions prone to unpredictable resource crashes, a lower expenditure of energy, made possible by a lower metabolic rate, would increase an animal's survival prospects in unfavourable times. In contrast, mammals that evolved in the highly seasonal, relatively predictable Arctic and Palaearctic environments have relatively high metabolic rates.[20] In other words, marsupials do not have a lower metabolic rate because they are marsupials, so the new theory implies, but because they have evolved in Australia.

In order to achieve these low rates of metabolism, various physiological and behavioural adaptations are needed. One that many small mammals employ is the ability to enter torpor. Found among many mammals in the strongly seasonal environments of the high latitudes, torpor is well known as a means of surviving intense cold. But recent studies have revealed that among many mammals in Australia and southern Africa at least, torpor is also common. But whereas the trigger for torpor among Holarctic mammals is the shortening day length, many of the mammals of the erratic environments enter torpor irrespective of the time of year. For them, the cue for torpor is food deprivation, because in these environments, the recurring hardship is not cold, but the unpredictable lack of rain.

The peculiar lifestyle of African mousebirds is one example of behavioural and physiological adaptations associated with a low metabolic rate. The mousebirds of sub-Saharan Africa have low metabolic rates in general, and to help them achieve such energy savings have evolved two very peculiar forms of behaviour. The first, known only among mousebirds, is their ability to cluster together at night into a tight ball. Those in the centre cling to the perch, and those on the periphery hang on to the pack. By clustering, the birds save up to 50 per cent of their nocturnal energy expenditure.[21]

Allied to this clustering behaviour is an unusual means of thermo-regulation. At night, the birds have been shown to drop their body temperature by as much as 14°C, then making optimum use of the sun the next day to regain their normal temperature. In a style known only to mousebirds, the birds hang vertically from their perches, using their erect tails as a brace to allow their bellies maximum exposure to the sun. Relative to other similar sized birds, the relatively low energy requirements of mousebirds enables them to eat less and utilise food sources of poor nutritional quality, such as acacia leaves.

Undoubtedly in the future many more marine and terrestrial species will be shown to have population numbers and life-histories that strongly fluctuate to El Niño's rhythms. Untangling the degree to which El Niño has influenced their evolution from other environmental factors might prove a much harder task. It is clear, however, that it has been a major evolutionary force.

9

A Forecast for Change
Predictions for an Unsettled Future

'Ow it can rain', the old man said, 'With things the way they are?
You've got to learn off ant and bee, and jackass and galah
And no man never saw it rain, for fifty years at least
Not when the blessed parakeets are flying to the east
The weeks went by, the squatter writes to tell his bank the news
'It's still as dry as dust' he said. 'I'm feeding all the ewes'
The overdraft would sink a ship, but make your mind at rest
It's all right now, the parakeets are flying to the west
                          Banjo Paterson, 'The Weather Prophet'.

As the Government Meteorologist of Queensland during
1902, Clement Wragge was under considerable pressure to
do something about the appalling drought that was gripping
the eastern half of his newly independent country. Australia
had already endured several years of below average rainfall,
especially during the El Niño of 1899–1900. But by 1902
conditions were even worse and half the country's livestock
was dead. In September that year, Wragge foolishly staked his
reputation on bringing the drought to an early conclusion.

A man of considerable energy and influence among meteorologists, Wragge was also known for his eccentric but
antagonistic manner. He frequently locked horns with his

peers and the authorities, earning himself the nickname of 'Inclement'. One of his most enduring contributions to meteorology was his idea to name tropical storms after people, often using names of politicians with whom he had clashed, comparing them to a cyclone 'whooping around and making a nuisance of itself'.[1] If the politicians threatened to sue, Wragge used their wives' names instead.

A subscriber to the unorthodox theory that gunfire could influence rainfall, Wragge believed he could trigger precipitation by discharging a series of Steiger Vortex cannons – 15-foot-long guns with cone shaped, vertical muzzles. At his behest, six cannons were sited at intervals of a mile near the outback town of Charleville, each to be fired every thirty seconds over a ten-minute period. The gala day chosen was 26 September , when the townsfolk came out in force to witness the spectacle. In the absence of Wragge on the big day, the Mayor of Charleville presided over the proceedings. The guns were duly fired. The drought continued.

Wragge blamed the operation's failure on the inadequate enactment of the experiment. He complained there should have been ten cannons instead of six, that only ten shots in total were fired and, to add insult to injury, two of the six guns had exploded hours prematurely through an overcharge of gunpowder. 'If the Charleville people will not carry out my instructions, I cannot help it,'[2] he told a reporter. However, his credibility with the public was shattered, and the press had a field day ridiculing his 'portentous prophecies'. When he was later overlooked for the job of Commonwealth Meteorologist, a position he had confidently expected to be granted, he left the country in disappointment, spending his final years in New Zealand.

Wragge of course had no idea that his confidence in his ability to end the drought flew in the face of unimaginably large forces operating in the Pacific. The futility of his actions may have become obvious had he paid more attention to the early historical accounts of drought in the outback. These might have told him that as a general rule-of-thumb when a drought is already under way in Australia by the middle of a year, it is highly unlikely to break until early into the following year. Wragge may also have benefited from reading *Rebellion in the Backlands*, written only a year beforehand by the renowned Brazilian novelist Euclides da Cunha. He observed that in the north-east of his country, 'the drought cycles . . . follow a rhythm in the opening and closing of their periods that is so obvious as to lead one to think that there must be some natural law behind it all, of which we are yet in ignorance'.[3] In 1902, these 'natural laws' were behaving as they have done for many millennia, and the first rains did not arrive in Australia until March 1903.

Meteorologists are no longer in total ignorance of the forces that baffled Wragge. There is now not only a broad understanding of the way in which the atmosphere and oceans work in tandem to drive the El Niño cycle, but scientists have infinitely more data at their fingertips to ascertain the state of the Pacific. Satellites with radar altimeters, sensitive enough to register fluctuations in sea-level of just a few centimetres, monitor subtle changes in the oceans every ten days. And spread across the Pacific is an array of sensor-laden buoys transmitting daily pictures of wind, atmospheric pressure, salinity and water temperature to a depth of 500 metres to orbiting satellites.

Beyond several days ahead, it becomes impossible to make meaningful forecasts about the day-to-day weather because random winds and eddies will always ensure the atmosphere's behaviour is unpredictable. But on a slightly longer time-scale, the behaviour of the atmosphere is controlled more by sea-surface temperatures. And since the oceans have a long memory, because of their large mass and heat capacity, knowledge of the behaviour of the ocean at the sea surface offers us a degree of climatic predictability. With such information, the potential exists to make forecasts several months in advance of the likely evolution of El Niño conditions. Perfect predictions are not possible, but reliable probability forecasts are. The behaviour of the atmosphere and oceans will always be affected by chaotic processes – or 'noise' as climatologists like to refer to it. Additionally, the initial state will never be perfectly known, no matter how detailed our ocean and atmospheric measurements can be. Predictions based on slightly varying initial states can lead to widely varying outcomes. To get around this, some models run an ensemble of predictions that are combined into a probabilistic forecast. The higher the consistency between each of the individual predictions in the ensemble, the higher the confidence in the overall forecast.

El Niño prediction has come a long way in a very short space of time. Consider that not only was the 1982–3 event, one of the two strongest of the century not predicted, it was hardly even noticed until it had reached its peak. Since then, increasingly complex forecasting models, the most sophisticated of which soak up hours of supercomputer time, have been developed. By 2005, scientists are confident that we will be able to accurately predict El Niños a year in advance in over 70 per cent of events – roughly the same probability of accuracy that currently exists for two-day weather forecasts.

Forecasts are currently made from two basic systems. The simplest are the statistical models, which search for analogues to the current conditions in comparable data over the last few decades. Their strength lies in their basis on real events, but the short period in which data have been collected, and the fact that no two El Niños are exactly the same, limit their application. The second broad type – the general circulation models, or GCMs – attempt to simulate the physical laws that govern the oceans and atmosphere, and the interactions between the two. Some are highly simplified, and attempt to represent only a few processes within a limited geographical range. An example is the GCM of the Lamont-Doherty Earth Observatory, containing a relatively simple representation of 'coupled' atmosphere and ocean just within the tropical Pacific, and which in 1986 passed the milestone of being the first to successfully predict an El Niño. Other GCMs are much more sophisticated, incorporating many parameters of the sea, atmosphere and land around the globe. It is believed that in the future, the best predictions will come from these global GCMs.

As yet, no one model has been recognised as being more likely than another to accurately forecast the next El Niño. Prior to the 1997–8 El Niño, there were over thirty institutions each with their own model. Most accurately predicted that an El Niño was about to take place, although some of that success was to be expected given the strength of the El Niño signal. On the other hand, very few spotted the event's main characteristic features – its intensity and the rapid speed of its development. Contrary to earlier successes, the Lamont-Doherty model did not perform so well, forecasting a mild warming for the Pacific at best. The most successful prediction came from a sophisticated global model developed by the European Centre for Medium Range Weather Forecasts, which projected that the event would be both very strong and would evolve rapidly in early 1997. But because at the time this model was still in the experimental stages of development, its conclusions were not placed in the public domain.

The potential users of reliable forecasts are many and varied, likely to save billions of pounds and thousands of lives. The insurance industry could adjust its pricing to adequately cover the changing risk burden. Emergency services could make important contingency plans. Fisheries bodies could adjust quotas or even shorten fishing seasons. Fire authorities could undertake protection burning programmes and firebreak maintenance. Governments could prepare relief programmes and initiate infrastructure projects such as dam and bridge reinforcement. President Fujimori of Peru was a very visible exponent of this proactive approach in early 1997, donning work clothes and assuming personal responsibility for civil defence works, doing his re-election chances no harm in the process.

The greatest financial benefits of accurate El Niño forecasts could be reaped by farmers. The use of El Niño prediction for agricultural purposes dates back hundreds of years. In order to know when to plant their staple potato crop, the Aymara people of the Bolivian and Peruvian Andes have for centuries made observations of Pleiades, a cluster of stars in the constellation Taurus. Six of the stars are visible to the naked eye, and together have figured prominently in the myths and agricultural cycles of many South American cultures. For several days leading up to the winter solstice festival of San Juan on 26 June, farmers wake up an hour or two before dawn to observe Pleiades above the horizon in the north-east. Special note is taken of the stars' brightness, size and the date of their first rising, the combination of attributes determining the amount of summer rain to be expected several months hence. The less visible the stars, or so the Aymara theory goes, the less rain will fall. Because the potato is most susceptible to drought when first establishing its roots, the later it is planted at these times the better. In the words of a pre-Hispanic chronicle, the Pleiades stars had much to reveal. 'If they come out . . . at their biggest people say, "this year we'll have plenty". But if they come out at their smallest, people say "we're in for a hard time".'[4]

There may be a very sound basis for this. It is known that in El Niño years, the rainy season tends to start late, and produce less rain, particularly in the agriculturally important months of December and January. Temperatures also tend to be warmer, creating an added negative effect on potato yield. It has also been established that in El Niños, the cloud layer in the upper atmosphere is significantly thicker – averaging 50 per cent cloud cover at an altitude of 15 kilometres as opposed to 35 per cent during La Niñas. This difference is estimated to be in the same order of magnitude as the change in the visibility of Pleiades.

These days, most relevant institutions extract their El Niño forecasts through international contacts and via the Internet, particularly monitoring the ongoing forecasts offered by the major climate-research institutions of the US, Europe and Australia. In most developing countries, this information is interpreted to establish a regional forecast, and via government warnings and media broadcasts is passed on down to farmers and communities. Awareness of the phenomenon is now high in virtually all countries strongly affected by El Niño, and warnings of an impending event may have significant global impacts on crop choice. In northern coastal Peru, an expectation of El Niño rains encourages a switch from cotton to rice, a crop less susceptible to water-logging than cotton. In Australia, the prospect of a drier year with later frosts makes chickpeas, with their shorter growing season and later planting date, less risky than wheat. In the corn-belt of the US Midwest, the economics of planting soya bean instead of maize are significantly improved with accurate La Niña forecasts. On the Argentine pampas, both maize and soya bean tend to be negatively affected during La Niñas. In these years, sunflowers, which demonstrate little response in terms of yields, become a viable alternative.

El Niño prediction offers great potential benefits to developing nations that rely on their agricultural sector for food security, employment and foreign revenue. Early awareness can lead to the importation of grain at lower prices and the provision of relief measures. It can also encourage farmers to plant food crops instead of cash crops. There have already been examples where advance warnings of impending El Niños have proved their worth in alleviating potential famine. In 1986–7, the National Meteorological Services Agency of Ethiopia used knowledge of developing El Niño conditions to issue warnings of drought for the *kiremt*, the rainy season between June and September. At the same time it predicted good rains for the *belg*, the short spring rains between February and May. Fearful of a repeat of the catastrophic famine of 1983–4, the government encouraged maximum effort during the *belg* by providing seed and fertiliser. It also promoted the adoption of conservative practices during the *kiremt* and was able to alert relief donors. The result was production increased by a third during the *belg*, and despite the severity of the subsequent *kiremt* drought, not a single life was reported to have been lost as a consequence.

Other countries have had mixed experiences in negotiating the pitfalls of El Niño prediction. In north-east Brazil in late 1991, following the issue of a forecast for drought, the governor of the state of Ceara toured the countryside, dispensing warnings and urging farmers to take appropriate action. Newspapers and radio stations joined in the governor's campaign, and the message was widely disseminated. Grain production fell by just 18 per cent of the mean, against the background of a 27 per cent reduction in average rainfall. This compared very favourably to the previous El Niño drought in 1987, when no forecasts were issued. In that year grain production plummeted by 85 per cent in response to a 30 per cent decline in average rainfall. The governor and FUNCEME, the state meteorological bureau that issued the forecast, received much warranted praise. However, their subsequent intervention has met with less success. Despite mixed indications that rainfall in 1993 would also be below normal, in late 1992 FUNCEME issued a forecast for normal rain. The prediction followed a certain amount of political pressure that a second consecutive drought forecast would be unfavourable during the governor's re-election campaign. When precipitation dropped by 40 per cent of the mean, and grain production declined by a similar amount, FUNCEME was widely discredited and has struggled ever since to restore its reputation.

During the 1997–8 El Niño, government organisations in Zimbabwe and Zambia were heavily criticised for erring in the opposite direction. As it became clear that an exceptionally strong El Niño was developing in the Pacific, severe drought warnings were issued and the media predicted the 'event of the century'. Farmers responded with drastic measures, culling livestock and planting drought-resistant but less profitable seed. Some farmers were too fearful to plant anything at all. At the same time, the authorities imposed water restrictions, demanded extra boreholes be sunk and prepared for food imports. In the event, an unusual warming of the tropical Indian Ocean mitigated the strength of the El Niño signal, and the summer rains were only slightly below average. However, the caution exercised by farmers in seed selection and that of banks in refusing credit led to an overall decline in grain yields of 40 per cent. As is often the case, the messenger was blamed, this time for what was perceived as unnecessary panic. In Zambia, there were even public calls for the forecasters 'to resign on moral grounds of disinformation'.[5]

It is the individual farmer who is ultimately most exposed to El Niño's capricious nature. He has the most to lose, but he also has an opportunity to profit by making the right decisions in the good years. Knowledge of impending conditions can be used to determine when to plant, what crop or varieties to select, which fertilisers and weed controls to employ and what stocking rates are most desirable for livestock.

However, in spite of the great potential to assist farmers in their planning processes, El Niño predictions are often of limited use at the village level. Many relate to broad regions and fail to be 'spatially specific'. In any case, the nature of probabilistic forecasts is not yet well understood, as the recent experiences in Brazil and southern Africa have shown. In future, more attention will be required to tailor forecasts to accommodate farmers' decision-making processes.

In Queensland, Australia, much success has already been achieved in this direction. A software program has been developed that provides statistical forecasts designed to meet individual needs. Special training is even available to help farmers understand the limitations of the forecasts and to make the best use of them. There has been a distinct shift away from seeing climatic anomalies as aberrations, and a trend towards viewing them as a natural part of the long-term cycle which must be managed for. The approach being fostered is for farmers to try to exploit the good years, but to pull their heads in during the bad. In the highly variable environment of Queensland, one study has shown that following that approach, 70–80 per cent of profits are now typically being made in the best three years of a decade.

While the accuracy of forecasting continues to improve, recent events suggest that the research goalposts may be shifting. There is little doubt that since the mid-1970s, El Niño has been behaving very strangely. In 1975, for example, all the usual indicators that an El Niño was evolving were present, but just as research vessels were dispatched into the Pacific to investigate them, the ocean and winds returned to normal. The 1982–3 El Niño event was exceptionally strong, certainly the strongest for over a hundred years, and was not expected to be rivalled by another for 100–400 years. Yet, the event fifteen years later achieved just that. In 1990, another event confounded expectations by failing to decay after the usual twelve months, continuing instead for another four years. Similarly, the 1998 La Niña did not decay as expected in early 1999, continuing into 2000. And overall since the mid-1970s, despite the recent extended La Niña, there have been far fewer such events than El Niños.

No one is certain how much of this aberrant behaviour can be attributed to man-made causes. Variations in El Niño behaviour have to be expected over time. Analysis of historical records, together with proxy data such as ice cores, reveals long-term fluctuations in the El Niño record. For example, there was less El Niño activity between 1925 and 1950 – apart from the strong 1939–41 event – than might have been expected. In addition, the linkage between an El Niño event and a particular regional response to it can vary, so that for certain periods of time, a region's weather is not affected in a typical fashion. However, many scientists doubt that this natural waxing and waning is enough to explain fully the highly unusual post-1975 behaviour. Based on a statistical analysis of the historical record, one study simulated a one-million-year data set, and concluded that the recent trend for more frequent and prolonged El Niños was likely to occur naturally only about once every 2,000 years. To many this suggests that El Niño's recent behaviour can best be explained by the influence of global warming.

We do know that over the twentieth century, global mean temperatures rose by between 0.3° and 0.9°C. Over this century, temperatures are expected to rise a further 2 ± 1°C. This would make the earth the warmest it has been for over 150,000 years. Global rainfall is also expected to change. The twentieth century witnessed a global rise of 2 per cent, with some parts of the mid-and-high latitudes receiving an extra 10 per cent. A further global increase of between 3 and 15 per cent is expected over the twenty-first century, although this will not be evenly distributed, and some regions may even see their annual precipitation fall. The trend for a greater percentage of rainfall to result from heavy downpours, as first observed over the nineteenth century, is expected to continue. More frequent floods and storms can be expected as a consequence. Sea-levels are also expected to rise by a further 50 ± 35 centimetres by AD 2100, having already risen approximately 10 centimetres during the previous hundred years. It is still unclear whether the number of cyclones, typhoons and hurricanes will rise, but it appears that there has already been a 40 per cent increase in intense tropical storm activity since 1969.

Studies investigating the likely impact of these changes on El Niño have so far tended to be contradictory and therefore somewhat inconclusive. Some models suggest that under increased CO_2 conditions, El Niño events will intensify and become more frequent. Others suggest that the amplitude of events will not alter, but there will be a permanent change in the 'mean state', so that what are known as El Niño conditions today will become much more the norm. There have even been some models that predict that the 'mean state' will shift in the opposite direction, so that the world will experience more permanent La Niña-like conditions.

The most profound problems to society will always relate, not to the gradual background rise of temperature and sea-level, but to the extremes. An increase in the frequency of floods and heavy rainfall will increase the rate of soil loss, and the damage to infrastructure. Most of the damage resulting from rising sea-levels will occur during violent storms. A greater incidence of drought will perpetuate the poverty of even more farmers. More outbreaks of large fires, such as those that occurred in 1997–8 in Indonesia, Brazil and Colombia, will increase carbon emissions.

There is already a widespread perception that we have begun to experience more extreme weather, particularly tropical storms, floods and tornadoes. This could be due partly to the increased frequency and intensity of El Niños during the last quarter of the twentieth century. It could also be attributed to the steeply rising costs associated with natural disasters in the Western world. What cannot be doubted is that there is now a much greater awareness of the impact of extreme weather events, fuelled by both their televisual power and the increasing anxiety over global warming.

It is also true to say that more people are becoming increasingly vulnerable to climatic extremes. As the best agricultural areas are already being tilled, population growth pushes ever more people into farming on marginal zones, usually in regions of poorer soils and highly variable climate. With them the newcomers bring agricultural techniques unsuited to such erratic rainfall – land clearance, slash-and-burn and cultivation methods – that will only increase the impact of extreme weather. Traditional systems of social insurance are disintegrating too, as more people are relying on the theoretical protection of national and international help. Even without an increase in the frequency of extreme climatic anomalies due to global warming, or alterations to the El Niño cycle, more people will be exposed to climatic vagaries.

There is a paradox here. On the one hand, the enormous advances that have been made by scientists in El Niño prediction allow millions of people in the developing world to significantly reduce the risks inherent in farming, and provide an opportunity to break free from the cycle of being victimised every time their growing season fails. On the other hand, more people are being shifted into precarious conditions, exposing them to even greater risks.

If recent El Niños are anything to go by, the great human tragedies and the social unrest that have occurred repeatedly throughout El Niños are set to continue. More people than ever will be affected, and the ripples of El Niño's impacts will travel further than ever before.

Notes and Sources

Introduction

1. *New York Times*, May 1907, in Gardiner and Van Der Vat p. 50.
2. Birch and Marko 1986.
3. The mechanism by which El Niño conditions in the Pacific actually influence iceberg behaviour is still unclear. It is thought that it may deepen the Icelandic low pressure system, thereby strengthening northerly winds over the Labrador Sea, which in turn drive more icebergs further south than normal. Moderate and weak El Niño events do not appear to affect the Icelandic Low.
4. The *Titanic* was following the Outward Southern Track, the designated shipping route for mail steamers during January to August. The route was selected to avoid ice and fog as much as possible, following a wide arc between Fastnet and 42°N, 47°W, then west to Nantucket Shoal light vessel, and on to New York.
5. As per Sailing Directions: United States Pilot (East Coast) 1909, in Parliament Great Britain, 1990.
6. Editors of Encyclopaedia Britannica 1978, lliffe 1990, Fisher 1927, Bhatia 1963, Maylam 1986, Lovejoy and Baier 1975, Luling 1975, Meggers 1994, Hugenholtz 1986, Barry 1997.
7. The ban lasted between September 1911 and January 1912, Hugenholtz 1986 p. 180.
8. Solomon and Stearns 1999.
9. Scott 1913 Vol. 1, p. 584, 606.

10. *New York Times*, 20 April 1912, p. 4.
11. US Coast Guard International Ice Patrol website.
12. Gardiner and van der Vat, p. 86.
13. Gardiner and van der Vat, pp. 85–6.
14. Lightoller, p. 222.
15. A letter by naturalist H. Tweedle, in Murphy 1926, p. 35.
16. Murphy 1926, pp. 32, 50, 51.

The Uncontrollable Child

Principal Sources: China – Arnold 1988, Aykroyd 1974, Bohr 1972, Evans 1945. **India** – Aykroyd 1974, Bhatia 1963, Presidency of Madras 1877, Seavoy 1986, Srivastava 1968, Stein 1998. **The Pacific** – Connell 1986, Lyons 1987. **South America** – Brooks 1971, Hall 1878, Loveman 1988, Smith 1880. **North America** – Bloom 1993, Diaz and McCabe 1999, Hoyt and Langbein 1955, Rousey 1985. **Africa** – Butzer 1976, Lindesay and Vogel 1990, Nicholson 1999, Walford 1879. **Australasia** – Dijken 1879, Potter 1993, Potter 1997, Rolls 1984. **El Niño history** – Allan et al 1996, Caviedes 1984, Endfield 1988, Glantz 1996, Normand 1959.

1. Russell, p. 184.
2. May 1877, in Bohr, pp. 44–5.
3. Hendrik Berlage, in Kiladis and Diaz 1986, p. 1035.
4. In Bohr, p. 83.
5. In Bohr, p. 93.
6. In Bohr, pp. 18-19.
7. Monsignor Louis Monagatta, in Bohr, p. 21.
8. See Chapter 4, 1640–1.
9. Chairman of Foreign Relief Committee of Tientsin, in Aykroyd, p. 81.
10. In Bohr, p. 7.
11. Letter to Richard Temple, January 1877, in Bhatia, p. 90.
12. In Presidency of Madras, 1877, p. 17.
13. In Connell, p. 62.
14. In Greenfield, p. 375.

15. Theophilo's História de Sêca, in Hall, p. 5.
16. In Walford, p. 36.
17. Memphis *Appeal*, 26 July 1879 in Bloom, p. 229.
18. In Bloom, p. 1.
19. In Walford, p. 19.
20. In Junker, p. 483.
21. In Walford, p. 49.
22. In Bock, p. 201.
23. In Potter 1997, p. 291.
24. Rolls, p. 183.
25. In Jenkins, pp. 121, 123.
26. In Beatson, p. 43.
27. In Todd, 1888.
28. See Chapter 4 for some of the global political and economic consequences of the 1972–3 event.
29. In Smyth, p. 430.
30. In Macculloch, Vol. 1, p. 225.

The Global Connection

Principal Sources: Physical processes – Allan et al 1996, Enfield 1989, Gribbin and Gribbin 1997, Salstein and Rosen 1984, Trenberth 1996. **Teleconnections** – Allan et al 1996, Basher and Zheng 1995, Chang and King 1994, Diaz and Kiladis 1995, Fraedrich 1994, Glantz et al 1991, Green et al 1997, Kiladis and Diaz 1989, LeComte 1989, Ropelewski and Halpert 1987, Trenberth et al 1988, Trenberth and Branstator 1992 Whetton 1997, Whetton et al 1990. **Exxon Valdez** – Cahill 1990, Keeble 1994, Wheelwright 1994.

1. Philander, p. 108.
2. In Thayer and Barber 1984.
3. In Eriksen 1993.
4. The calculation was based on the heat content above the main thermocline for the area within 5°S–5°N, 120°E–80°W. Between April 1997 and August 1998, the loss in warm water

volume to the higher latitudes corresponded to 7.5e + 22 Joules, or the equivalent of 900,000 20-Mt bombs. During the build-up of warm water in the region between September 1995 and April 1997, which encompassed the 1995–6 La Niña, the heat content of the region increased by 3.8e + 22 Joules, or the equivalent of 450,000 20-Mt bombs. (Chris Meinen and Billy Kessler, N.O.A.A. pers. comm.)

5. In Nicholls 1988.
6. Caviedes 1991, Whetton and Rutherford 1994.
7. Editors of Encyclopaedia Britannica 1978.

The Art of Survival

Principal sources: Australia – Blainey 1983, Bowman 1931, Flannery 1994, Heathcote 1969, Nicholls 1988, Roderick 1991, Ward 1980. **Inter-community Alliances** – Brookfield and Allen 1989, Cashdan 1980, Cashdan 1986, Cashdan 1989, Colson 1979, Flannery 1994, Hayden 1975, Hayden 1992b, Hayden 1992a, Hayden 1993, Hitchcock 1979, Latz 1995, Lee 1968, Lee 1976, Minc 1986, Nicholls 1989b, Peterson and Long 1986, Piddocke 1976, Strehlow 1965, Suttles 1987, Waddell 1975, Waddell 1989, Wiessner 1980. **Belief systems** – Arnold 1988, Bell 1971, Bell 1975, Eberhard 1977, Elvin 1998, Fagan 1996, Fagan 1999, Feng 1976, Hassan 1994, Meyer 1984, Popper 1951, Said 1993, Sinclair 1987, Tuan 1980. **Sacrifices** – Bourget (in press), Dening 1980, Kolata 1996, LaBarre 1943, Ortiz de Montellano 1978, Redfield and Villa Rojas 1962. **Christianity** – Curcio 1997, Gibson 1964, Gill 1871, O'Hara and Metcalfe 1995, Tai 1983, Tarling 1992.

1. *Bourke* in Lawson 1979.
2. See Chapter 4, 1891–2.
3. *In defense of the bush*, Paterson 1993.
4. In Musgrave, p. 26.
5. In Mitchell 1838, Vol. 2 p. 13.
6. In Oxley, p. 246.
7. In Sturt 1833, Vol. 1 pp. 48, 73.

8. In McKinlay, p. 232.

9. In Johnston, p. 300.

10. In Heathcote 1987.

11. *Bourke* in Lawson 1979.

12. Flannery 1994, p. 392.

13. Chatwin, p. 16.

14. Wiessner, p. 77.

15. As said to a lay missionary in Laiagam, 1972, in Waddell, p. 251.

16. Tawney, p. 77.

17. *Antony and Cleopatra*, II vii, 19–26, M. Spevack edition 1990 pp. 134–5.

18. J. Atens, in Said, p. 96.

19. Litchtheim, Vol. 1, p. 206

20. Snow, p. 457.

21. In Williams 1899, p. 467.

22. Yen King-Ming, Special High Commissioner for the Superintendence of the Arrangements for Famine Relief in Shanxi, in Walford 1878, pp. 18-19.

23. Meek, p. 130. Jukun was a powerful Sudanic kingdom of the late European Middle Ages, centred around northern Nigeria.

24. Chapman 1982, in Loeb 1971.

25. Aldred, p. 158.

26. King Amenemhet of the Eighth Dynasty, in Bell 1971, p. 18.

27. In Aldred, p. 161.

28. Inscription on Ankhtifi's tomb south of Luxor, in Bell 1975, p. 9.

29. Missionary William Crook, in Dening, p. 102.

30. For discussion re Aztec sacrifice numbers, see Ortiz de Montellano.

31. Rougeyron 1849, p. 9.

32. March 1846, The Reverend William Gill, Vol 1, p. 163.

33. Macgregor 1937.

34. 15 March, 1846, Rarotonga, as in Gill, pp. 87–91.

35. Protestant missionary Mikael Aragawi, in Pankhurst 1985, p. 109.

36. Lazarist missionary Father Crombette, in Pankhurst 1985, p. 109.
37. In Bohr, p. 102.
38. In Bohr, p. 124.
39. January 1878, in Tai, p. 50.

Reign of Terror

Principal sources: Peruvian history – Donnan 1990, Moseley 1987, Moseley 1992, Moseley 1999, Moseley et al 1981, Moseley et al 1992, Nials et al 1979, Richardson 1994, Shimada et al 1991. **Mega-niños and cultural responses elsewhere** – Arnold 1988, Binford 1997, Hodell 1995, Meggers 1994, Meggers 1995, Rotberg and Rabb 1983, Wilkerson 1994. **Chronology** – Hamilton and Garcia 1986, Quinn 1992, Quinn and Neal 1987. Thompson et al 1984, Thompson et al 1986, Whetton and Rutherford 1994. **Egypt 1200** – Queller 1997, Watterson 1997, Zand et al 1964. **Aztecs 1450–4** – Brundage 1979, Davies 1977, Hassig 1981, Katz 1972, Kovar 1970, Soustelle 1964, Townsend 1992. **China 1640–1** – Chan 1982, Dunstan 1975, Mote and Twitchett 1988, Parsons 1970, Spence 1978. **Bengal 1769–70 – Ghose 1943, Hunter 1897, Spear 1965. Ireland 1844–6** – Aykroyd 1974, Bourke and Lamb 1993, Lamb 1995, Smith 1933. **Ethiopia 1888** – Pankhurst 1985, Zewde 1976. **Russia 1891–2** – Figes 1996, Robbins 1975, Wilson 1988. **Vietnam 1930–1** – Buttinger 1967, Duiker 1996, Du Bois 1974, Scott 1977. **The world 1972–3** Coetzee 1974, Glantz 1996, Holt and Seaman 1976, Lamb 1995, Legum 1975, Marcus 1994, Ziring 1981. **Mexico** – Gibson 1964, Hart 1989, Meyer 1984, O'Hara and Metcalfe 1995, Swan 1982, Swan 1995. **South Africa** – Ballard 1986, Eldredge 1992, Giliomee 1979, lliffe 1995, Laband 1997, Laband and Knight 1996, Lindesay and Vogel 1990, Milton 1983, Morris 1965, Mostert 1992, Peires 1982, Shillington 1989, Venter 1985, Vogel 1989, Walker 1928. **Slavery** – Arnold 1988, Bennett 1976, Binder 1977, Brooks 1971, Conrad 1993, Dias 1981, McDonald 1982, Miller 1982, Miller 1988, Tinker 1974.

1. In Cohen 1997, p. 85.
2. The earthquake was 20 February 1835. Darwin sailed into Concepción 5 March 1835. Darwin 1933, pp. 280–1.
3. In Arnold 1988, p. 31.
4. In Arnold, 1988, p. 31.
5. Excavation work by Christopher Donnan at the ruins at Chotuna – centre of Naymlap Dynasty, dates Naymlap's arrival at AD 750, and end at AD 1100, when there is evidence of flooding.
6. Famines were also experienced during 1200–1 in India and China.
7. Zand et al, 1964, pp. 225, 227, 229, 245, 273.
8. Reinforcing the evidence that 1450–4 were El Niño in La Niña years, the Nile was exceptionally low in 1450 (at Bulaq, people could wade between the shorelines), while there was also famine in northern China. Two years on, mirroring the switch in extremes that happened after two years of drought in Mexico, 1452 saw serious flooding of the yellow and Huai Rivers. Heavy rains and severe cold winters continued in northern China for another two years, to the extent that in 1453–4, the sea mouth of the Huai River froze over.
9. Fernando de Alva Ixtlilxochitl, in Kovar, p. 28.
10. Part of royal supplications to Tlaloc, in Brundage, p. 72.
11. Relations of Chalco Amaquemecan, in Kovar, p. 31.
12. In Duran, p. 147.
13. In Bancroft 1876, Vol. 5, p. 414.
14. In Townsend, p. 200.
15. Tlacaellel, Chief Deputy Speaker of the Aztecs, in Katz, p. 169.
16. In Japan, the period 1640–1 became known as the *Kan'el famine*, after appalling weather led to almost total harvest failure. There was also drought in Vietnam, Mexico, and heavy flooding in central Africa.
17. *Suining zhigao* (history of Suining), in Wakeman, Vol. 1, p. 7.
18. In Spence, p. 4.
19. Essayist P'u Sungling, in Spence, pp. 20–1.
20. Dunstan 1975.

21. Although the year 1769–70 was not originally listed as an el Niño by Quinn, other researchers have since presented evidence that it was. There was flooding in California, eastern USA and northern Peru. Low rainfall was recorded in north China and Java in 1769, and the annual floods of the Nile were well below average. There was also drought in Chile, central Mexico and central Africa. And the winter in Europe and western Russia was exceptionally long and cold. The north and east coasts of Iceland were surrounded by sea-ice until the Autumn.

22. Report on the famine by Warren Hastings 1772, in Hunter, p. 381.

23. Macaulay, pp. 82–3

24. In Ghose 1943, p. 43.

25. Teignmouth p. 25.

26. Letter from the President to the E.I.C.'s Court of Directors, 18 March 1770, in Hunter, p. 27.

27. In Hunter, p. 27.

28. Hastings, in Hunter, p. 381.

29. Elsewhere in the world, during 1844–6, there were heavy rains in Peru, record floods in northern China, on the Mississippi River, and across western Europe, and several severe cyclones across central Polynesia. There were drought-induced famines in India, Siam, Australia, Russia, Borneo, Java, Hawaii and north-east Brazil.

30. Birmingham, 1993.

31. Dr O'Bryen Bellingham in Nelson, p. 6.

32. In Smith, p. 147.

33. Elsewhere there were also droughts in large parts of Australia, India, Sudan and north-east Brazil.

34. In Pankhurst 1985, p. 84.

35. In Pankhurst 1985, p. 79

36. In Pankhurst 1985, pp. 70, 109.

37. Elsewhere in 1891–2, there were exceptionally heavy rains in Peru, while in Shanghai, the mouth of the Yangtze river as well as the harbour were frozen for over a month.

38. Tolstoy to fellow writer Leskov, in Wilson 1988, p. 398.
39. Chekhov to his wife, in Figes 1996, p. 160.
40. Also in 1930–1, there was drought in north-east Brazil, Indonesia and eastern South Africa, and a terrible harvest in Japan. In Niger, 10 per cent of the population died from the combination of drought and locusts. Near the Sudan/Uganda border, the White Nile dried up, allowing Sudanese to walk over to the Madi to purchase food. In July 1931, over 70,000 square miles of central and north-eastern China were inundated after six gigantic flood waves washed down the Yangtze and Huai rivers. The death toll was estimated as at least 300,000, and between twenty-eight to forty million were made homeless.
41. Inspector of Sanitation and Medical Services, in Scott 1977, p. 138.
42. R. P. Derribes, in Scott 1977, p. 142.
43. Gov. Gen. Pasqurer, in Buttinger Vol. 1, p. 220.
44. For other El Niño-related impacts of 1972–3, see chapter.
45. In Kapuściński, p. 111.
46. Balandier, p. 39.
47. Drought and poor harvests were also experienced in Japan and the West Indies during 1785–6. Drought was so severe in the upper reaches of northern China's Huai river catchment that waters were too low to prohibit silt from the Yellow river accumulating in the Grand Canal, ruining that important trading route for decades.
48. The manager of the Tulancalo *hacienda* in central Mexico, 4 August 1910, in Swan 1995, p. 123.
49. Guadalupe Castorena, in Swan, pp. 12–13.
50. James Read, 4 June 1846, in Mostert, 1992, p. 871.
51. Grahamstown Journal, 15 May 1878, in Milton, p. 258.
52. Zand et al 1964, p. 247
53. Harms 1987.
54. Barbosa 1918, p. 125.
55. Annals of the Sao Paolo Assembly 1880, in Conrad, p. 123.
56. In Tinker 1974, p. 119.

57. In Moors, p. 271.
58. Estimates quoted by Henry Maude, in Fosberg 1970.

Uncharted Territory

Principal Sources: Pacific exploration – Bridgman 1983, Caviedes and Waylen 1993, Ferdon 1993, Finney 1985, Finney 1994, Irwin 1992, Levinson et al 1972, Parsonon 1972. **Magellan 1520–1** – Joyner 1992, Morison 1974, Nunn 1934 Obregon 1980, Pigafetta 1969. **Pizarro 1532** – Cohen 1981, Hemming 1983, Prescott 1847. **Marcos de Niza 1539–41** – Bandelier 1981, Day 1940, Hallenbeck 1949. **Drake 1578–9** – Wilson 1989, Wilson 1977. **1587–9** – Chang 1997, Douglas 1958, Dunstan 1975, Hall 1978, Herring 1979, Quinn 1974, Quinn 1984, Quinn 1985, Sargent 1979, Stahle et al 1998a, Wakeman 1985, Webster 1979, Webster 1980, Webster 1983. **Cape Town 1652** – Riebeeck 1952, Theal 1907. **1769–70** Hittell 1855, Dellenbaugh 1905, Crespi 1971, Blainey 1983, Gilbert 1981. **1791–2** – Spate 1988, Connell 1986, Douglas 1970, Billardiere 1800, Sparrman 1953, Lyons 1986, Vancouver 1984, Endfield and O'Hara 1997. **1812** – Palmer 1997, McLynn 1997, Perry 1963. **1845** – Tolcher 1986, Sturt 1849, Joy 1964, Livingstone 1961, Ransford 1978, Jeal 1973. **1888** – Stevenson 1892, Andrade 1981, Sprout and Sprout 1946. **1891–2**– Kimberly 1897, Macdonald 1996, Cannon 1987, Lindsay 1893. **1941–2** – Neumann and Flohn 1987, Neumann 1992, Stolfi 1980, Lucas 1998, Hoyt and Langbein 1955, Belden 1939, Cox and Large 1960.

1. Williams 1837, p. 45
2. Cook 1785, Vol. 2, p. 250.
3. In Heyerdahl, p. 31.
4. In Finney 1985.
5. In Finney 1994.
6. In Ferdon 1993.
7. Ward and Brookfield 1992, Estrada and Meggers 1961, Langdon 1993, and Langdon 1989. Other possible routes of contact between Asia and the Americas are via the coastal

route of the northern Pacific, or using the Pacific equatorial countercurrent.

8. In Beechey 1831.
9. In Arthonioz 1952.
10. In Levinson et al 1972.
11. In Cook 1785, Vol. 3, p. 26.
12. Corroborating the Magellan evidence that 1520–1 was an El Niño period is that the Nile's floods were extremely poor in 1520, so much so that to improve the inundation, special prayers were read. There was also a devastating flood of the Yangtze river, and in 1521, a famine was recorded in the Indian provinces of Sind. Harvests across much of Europe were also terrible, so bad that in Portugal, the year 1521 became known as 'The Year of Great Hunger'.
13. In Pigafetta, Vol. 1, p. 57.
14. Towle, p. 88.
15. Historian Antonio Herrera, in Nunn, p. 618.
16. In Joyner, p. 168.
17. In Obregon 1980.
18. In Pigafetta Vol. 1, p. 57.
19. Cohen, 1981, p. 71. Others have argued that San Miguel may have been situated on the perennial Chira river, so mention of flowing water would not be out of the ordinary, and therefore Quinn's assessment of 1532 as an El Niño is dubious. (Hamilton and Garcia 1986).
20. In Cohen, p. 77.
21. In Hallenbeck, p. 195.
22. In Hallenbeck, p. 34.
23. Meaning 300 mounted soldiers, in Hallenbeck, p. 25. Probably referring to the Sonora Valley.
24. In Day 1940, p. 112.
25. In Day, p. 114.
26. In Day, p. 114.
27. In Pankhurst 1985, p. 33.
28. In Pankhurst 1967, p. 72.
29. In Goodrich 1976, Mote and Twitchett 1988.

30. Bishop Marroquìn, as cited in Lutz, p. 7.
31. In Hanna, pp. 364–5.
32. In Hanna, p. 366.
33. In Hanna, p. 366.
34. In Bancroft 1884, pp. 144–5.
35. In Quinn 1974, p. 439.
36. In Stahle et al 1998.
37. In Walford, p. 11.
38. The Bishop of Skalhort, in Ogilvie, p. 107.
39. In Sargent, p. 241.
40. Friar João dos Santos, in Theal 1902, Vol. 7, p. 197.
41. Webster 1979, p. 12.
42. In Webster, Vol. 1 1980, p. 77.
43. In Wakeman, p. 107.
44. Elsewhere in 1652 there was drought in Guatemala, floods further north in Mexico, and drought followed by floods in the Yangtze river valley.
45. In Riebeeck, Vol. 1, pp. 45–6.
46. For other El Niño effects in 1769–70, see Chapter 4.
47. Cook 1793, p. 103.
48. Banks, Vol. 2, p. 57.
49. Blainey 1983, p. 12. Blainey cites four distinct reasons for the choice of settlement. Besides the over-optimism on the climate, there was the search for new naval supplies, the solution to the convict problem and a restocking station for trading routes.
50. 24 November, 1770, Banks, Vol. 2, p. 192.
51. Blainey 1983, p. 12.
52. It has been suggested by historian Richard Grove that the unusually severe weather and subsequent poor harvests of 1788–9 that helped precipitate the French revolution were associated with a 'precursor' to the exceptionally strong el Niño that began in late 1790. If it could be proven that this earlier period was actually a La Niña, this theory would certainly be given more weight.
53. In Beaglehole, Vol. 2, pp. 539, 541.
54. Cook 1777, Vol. 2, pp. 108, 118, 120.

55. Forster, Vol. 2, pp. 398–9.
56. Cook 1777, Vol. 2, p. 127. La Billardiere, Vol. 2, p. 200.
57. Sparrman 1953, p. 168. Although he visited the island with Cook, Sparrman was able to comment on the French experiences in New Caledonia as he did not write his journal of Cook's expedition until several years after D'Entrecasteaux's visit.
58. La Billardiere, Vol. 2, p. 213.
59. In Vancouver, Vol. 3, p. 895.
60. In Vancouver, Vol. 3, pp. 894–5.
61. In Vancouver, Vol. 3, p. 896.
62. In Beatson, p. 42.
63. In a letter to the Colonial Secretary, 4 March 1791, in Nicholls, 1988, p. 4.
64. In Spruce 1864, p. 29.
65. Lieutenant-Colonel Graham, in Peires, p. 65.
66. February 1812, Humboldt, Vol. 4, pp. 11–12.
67. In Etheridge, p. 36.
68. Collins's 'Account of the English Colony in NSW', in Perry 1963, p. 27.
69. Gov. Macquarie, in Blaxland, p. 50.
70. In Palmer, p. 252.
71. In Sturt 1849, Vol. 1, p. 305–6.
72. In Sturt 1849, Vol. 2, p. 86.
73. In Sturt 1849, Vol. 2, p. 21.
74. In Sturt 1849, Vol. 2, p. 89.
75. In Livingstone 1961, p. 103.
76. In Livingstone 1961, p. 63.
77. In Livingstone 1960, p. 301.
78. In Stevenson, p. 246.
79. In Stevenson, pp. 250–1.
80. In Stevenson, p. 267.
81. *New York Herald*, 31 March 1889, in Sprout 1996, p. 207.
82. *New York Times*, 5 March 1889, in Andrade, 1981, p. 79.
83. Royal Geographic Society of South Australia enquiry, in Macdonald, p. 22.

84. Lindsay and Wells, p. 182.

85. Lindsay and Wells, p. 147.

86. The exceptionally cold winters experienced throughout much of Europe between 1939 and 1942 had already had a major impact on the outcome of the war. The German offensive against Britain planned for late 1939 was progressively postponed until spring 1940 because of continuous cold, fogs, and snowfalls, allowing the British Fighter Command crucial time to strengthen its forces.

87. Operation Barbarossa, which began on the 22 June, was originally planned for 15 May, but delayed partly because of exceptionally heavy spring rains.

88. R. Scherhag, in Neumann and Flohn 1987, p. 627.

89. Hitler's instructions 8 December 1941, in Neumann and Flohn 1987, p. 626.

90. In Neumann and Flohn 1987, p. 625.

The Price to Pay

Principal sources: Indonesia – Ballard 1998, Brookfield et al 1995, Forrester and May 1998, Potter 1997, Schweithelm 1998, Vidal 1997, W.W.F. International 1997, 1997. **Modern Impacts** – Aust. B.O.M. 1985, Bouma et al 1997, Bryant 1991, Cane et al 1994, Changnon 1996, Dilley and Heyman 1995, Guillory 1996, Heathcote 1988, Smith and Ward 1998, Tiffen and Mulele 1994. **Human health** – Baylis et al 1999, Bouma and Dye 1997, Bouma et al 1996, Christophers 1911, Epstein 1997 Hales et al 1996, Kovats et al 1999 Lewins et al 1994, Marshall 1993, Nicholls 1993, Patz et al 1996, Wenzel 1994.

1. Du Bois, pt. 3, pp. 4–5.

2. In Kyodo News, 29 September 1997.

3. Ramon and Wall, in Schweithelm 1998.

4. Estimate by Willie Smits, (pers. comm.). The total population before the fires was estimated at 15,000.

5. There was not necessarily anything abnormal about the

weather patterns that produced the summer rainfall. The real anomaly, and greatest probable contribution of El Niño, was the extended period over which there was heavy rainfall.

6. In Frampton et al, p. 38.
7. Christophers, p. 9.
8. Gill 1921, p. 636.

Troubled Waters

Principal Sources: Peru – Arntz 1986, Arntz et al 1988, Arntz et al 1991, Barber 1988, Barber 1990, Barber and Chavez 1983, Barber and Chavez 1986, Duffy et al 1998, Idyll 1972, Laws 1997, Moseley 1975, Sharp and McLain 1993, Wilson 1981. **Coral** – Birkeland 1989, Buddemeir and Fautin 1993, Bunkley-Williams and Williams 1990, Davidson 1998, Eakin 1988, Eakin 1992, Eakin 1996, Rowan 1997, Wellington and Victor 1985, Wilkinson 1999. **Spondylus** – Marcos 1977, Marcos 1998, Paulsen 1974, Richardson 1994, Salomon and Urioste 1991. **Southern Ocean** – Brierley and Reid 1999, Croxall et al 1999, Heywood et al 1985, Laws 1985, Priddle 1988. **Fisheries** – Caputi et al 1996, Lehodey et al 1997, Lenanton 1991, Love 1987, Mysak 1986, Pearce and Phillips 1994, Pearce and Phillips 1988, Vance and Staples 1985, White and Downton 1991. **Green turtles** – Limpus 1989, Limpus and Nicholls 1988.

1. Heyerdahl 1995, p. 209.
2. Others have argued that the periodic downturn in food supply during El Niño events would make the development of civilisation on a purely maritime basis unlikely. (Wilson 1981.)
3. Cieza de Leon, p. 337.
4. In Forbes, p. 414.
5. In Murphy 1925.
6. Other species of damselfish have been observed keeping a similar check on urchin bioerosion in the Galápagos (Glynn 1988).
7. Leon Zann, Southern Cross University (pers. comm.).

8. In Salomon and Urioste 1991, p. 116.
9. Possible relationships have also been identified between El Niño and the reproductive activity and breeding success of several other bird and mammal species of the Southern Ocean, such as Crabeater seals *Lobodon carcinophagus*, Leopard seals *Hydrurga leptonyx*, Weddell seals *Leptonychotes weddellii* and Snow Petrels *Pagodroma nivea.. (Testa 1991, Chastel 1992). However, the physical linkages that underpin these relationships have not yet been demonstrated, and so El Niño's effects on the Southern Ocean as yet can still only be deemed circumstantial.*
10. The mechanism by which sea-ice cover determines krill population success is still not yet proven.
11. Refering to Nonouti Island, in discussion in Fosberg 1970.
12. The tendency of salmon to shift north into Canadian waters (also to the benefit of Alaskan fishermen) has in the past been the subject of a heated international dispute. Normally a significant proportion of returning Sockeye salmon travel south of Vancouver Island through the Juan de Fuca Strait, effectively allowing both Canadian and American fishermen a share of the bounty. In strong El Niños, waters around Vancouver Island tend to be warmer, causing most returning salmon to pass north through Johnstone Strait, effectively bypassing American fishermen.
13. One of the reasons proposed for the scallop's success is that it is likely to have had tropical origins, and despite having since adapted to cold water conditions, has retained a resistance or tolerance to warm temperature because of the adaptive benefits during El Niños.
14. The influence might be twofold: the warmer temperature of the current may improve growth and survival of the larvae, and may act to physically transport the larvae, enabling them to catch the fast track to the coast.
15. Fluctuations in the strength of the Leeuwin current have also been shown to correlate to several other important fisheries' species, such as the scallop *Amusium balloti* and western king

prawns *Penaeus latisulcatus* in Shark Bay, pilchards *Sardinops sagax neopilchardus*, Australian salmon *Arripis truttaceus,* Australian herring *Arripis georgianus* and whitebait *Hyperlophus vittatus.*

An Evolutionary Force

Galápagos – Gibbs and Grant 1987, Grant and Grant 1987, Grant and Grant 1993, Laurie 1989, Merlen 1984, Merlen 1985, Rechten 1985, Trillmich 1985, Trillmich 1987, Trillmich 1991, Valle and Coulter 1987, Weiner 1994, Wikelski 1997. **Ecosystem changes** – Wallace and Temple 1988, Myres 1988. **Springboks** – Cronwright-Schreiner 1925, Green 1955, Lovegrove 1993, Scully 1898, Skinner 1993. **Mass flowering** – Appanah 1993, Ashton et al 1988, Caldecott and Caldecott 1985, Dove 1993, Dove and Kammen 1997, LaFrankie and Chan 1991, Pfeffer and Caldecott 1986, Sakai et al 1999, Yasuda et al 1999. **Population distribution** – Berger 1998, Desender et al 1992, Dillon and Rundel 1989, Duffy 1989, Guerra and Portflitt 1991, Osborne and Davis 1997, Osborne 1989, Peck 1996, Pounds et al 1999, Pounds and Crump 1994, Richmond 1989, Trillmich 1993, Vermeij 1989, Zabel and Taggart 1989. **Adaptation** – Diamond 1985, Dickman et al 1995, Flannery 1994, Hughen et al 1999, Keast 1981, Lovegrove 1996, Low 1978, Morton 1986, Nicholls 1989a, Pearce 1999, Philipps 1990, Predavec and Dickman 1994, Schreiber 1994, Schreiber and Schreiber 1989, Snow and Nelson 1984, Vandenbeld 1988, Wikelski and Thom 2000, Wills 1996, Zann et al 1995.

1. Llosa, p. 17.
2. Coastal condors on the Sechura Peninsula did not breed during the 1982–3 El Niño as witnessed during the study of Wallace and Temple 1988. However, because of the particular pattern of currents during the 1982–3 event, dead marine animals that would normally be washed onto the Peninsula's beaches were, on this occasion, carried further south.
3. Some seasonal migration into the US occurs each year. Only in

certain years such as 1952, 1958, 1973, 1979 and 1983 do they reach western Canada in considerable numbers. There also appears to be some correlation in annual population figures between North America and Europe.

4. From an unpublished doctoral thesis of Chan Hung Tuck, University of Edinburgh.

5. The toll covers one whole mast season. There is disagreement as to how the spirits exact this toll, but the literal interpretation is that the spirits cause people to become ill and die.

6. The Boers divided the springbok into two categories, the term *trekbokken*, referring to those springbok of marginal lands that were subject to the periodic migrations, and the *houbokken*, which remained permanently on the veldt.

7. Cronwright-Schreiner, pp. 41, 49–50.

8. Dr Gibbons, in Green, p. 46.

9. The tale of Gert van der Merwe on the Molopo river, in Green, p. 37.

10. William Scully, magistrate of Springbokfontein, in Green, p. 43.

11. In Cronwright-Schreiner, p. 59.

12. In Darwin 1859, p. 84.

13. As a result of El Niño ocean conditions, there is often a lack of concordance between species' breeding performances at colonies which can generally be explained by the species' respective diets and feeding styles. In this case however it is unclear – kittiwakes are surface-feeders, for instance, but both murres and puffins are divers. It should also be noted that another study on Alaskan seabirds at the same time also found widespread breeding failure among kittiwakes, but great success among the common murres. (Hatch 1987).

14. From as yet unpublished work by Terry Erwin, Smithsonian Institute (pers. comm.).

15. The 1982–3 El Niño was documented as having resulted in the probable extinction of one species of hydrocoral, *Millepora sp.* (Glynn and Weerdt 1991).

16. Most temperate passerine birds, for example, breed in their

second year, lay 3–10 eggs per clutch, and have an average lifespan of a few years.

17. It may be more than coincidental that many of the formative seabird studies on which this concept was based happened to have been conducted in El Niño years (Schreiber and Schreiber 1989).

18. The banded stilts that breed on the salt lakes of central Australia, having travelled from virtually anywhere around much of the continent's periphery, appear to be a separate population from those that restrict their breeding to West Australia's lakes. Rainfall variability in the west shows a weaker relationship to El Niño and La Niña years, and a stronger correlation to variability in Indian Ocean sea-surface temperatures, than for the centre and east of the continent.

19. Breeding numbers of several species of waterfowl in Australia have been shown to correlate to El Niño fluctuations. The ephemeral wetlands of the interior act as important breeding sites, while the more permanent wetlands of the continent's periphery used by the waterfowl in non-El Niño years act mainly as refuges. (Norman and Nicholls 1991).

20. Wills, pp. 163, 168

21. Besides the El Niño cycle, aridity is also another source of resource unpredictability, and the relatively low metabolic rates of mammals found in desert regions tend to mimic the pattern of those found in regions strongly influenced by El Niño.

22. Barry Lovegrove and Andrew McKechnie, University of Pietermaritzburg, pers. comm.

A Forecast for Change

Wragge – Gibbs 1998, **Forecasting and Modelling** – Orlove et al 2000, Kerr 1998, Latif 1994, **Application of predictions** – Orlove and Tosteson 1999. **El Niño's future** – Enfield 1988, Glantz 1994, Knutson and Manabe 1994, Meehl and Washington 1996, Saunders 1998, Trenberth and Hoar 1996a & b.

1. In Gibbs 1998, p. 16.
2. Interview with Wragge, *Brisbane Courier* 29 September 1902, p. 6.
3. Cunha 1943, pp. 25–6.
4. In Salomon and Urioste 1991, p. 133.
5. In *Saturday Star*, Johannesburg 1 August 98.

El Niño, La Niña Chronology

The source for years 1870–2000 was the UK Meteorological Office, based on an analysis of sea-surface temperature data for the eastern equatorial Pacific. The chronology for the years 1525–1870 was based on Quinn 1992. Additional evidence for other events can be found, among others, in Hamilton and Garcia 1986, Whetton and Rutherford 1994, Konnen et al 1998 Dunbar 1994, Herring 1979, Kaplan 1998, Meyerson 2000, Stahle 1998b.

Bibliography

Aldred, C., 1961. *The Egyptians.*

Allan, R., J. Lindesay and D. Parker, 1996. *El Niño Southern Oscillation and climatic variability.*

Andrade, E., 1981. 'The great Samoan hurricane of 1899.' Naval War College Rev. 25: 73–81.

Appadorai, A., 1936. *Economic conditions in Southern India, AD 1000–1500.*

Appanah, S., 1993. 'Mass Flowering of dipterocarp forests in the aseasonal tropics.' J. Biosc. 18(4): 457–74.

Arnold, D., 1988. *Famine: social crisis and historical change.*

Arntz, W., W. G. Pearcy and F. Trillmich, 1991. 'Biological Consequences of the 1982–83 El Niño in the Eastern Pacific', in Trillmich, F. and K. A. Ono (eds). *Pinnipeds and El Niño: responses to environmental stress.*

Arntz, W. E., 1986 'The two faces of El Niño' 1982–83' Meeresforchung 31: 1–46.

Arntz, W.E., E. Valdivia and J. Zeballos, 1988. 'Impact of El Niño 1982–83 on the commercially exploited invertebrates of the Peruvian shore.' Meeresforsch. 32: 3–22.

Arthonioz, P., 1952. Une nouvelle expédition Kon Tiki: De Bikini aux Hébrides d'un Voilier Perdu en Mer.' J. de la Société des Océanistes 8: 295–8.

Ashton, P. S., T. J. Givnish, and S. Appanah, 1988. 'Staggered flowering in the Dipterocarpaceae: new insights into floral induction and the evolution of mast fruiting in the aseasonal tropics.' Am. Nat. 132(1): 44–66.

Australian Bureau of Meteorology, 1985. 'Report on the meteorological aspects of the Ash Wednesday fires', 16 February 1983.

Aykroyd, W. R., 1974. *The conquest of famine*.

Balandier, G., 1972. *Political anthropology*.

Ballard, C., 1986. 'Drought and economic distress: South Africa in the 1880's.' J. Interdisc. Hist. 17: 359–78.

Ballard, C., 1998. Political Reviews, Melanesia – Irian Jaya. *The Contemporary Pacific* 10(2): 433–40.

Bancroft, H.H., 1876. *The native races of the Pacific states of North America*.

Bancroft, H.H., 1884. *History of the north-west coast*. Vol. 1 1543–1800.

Bandelier, A.F.A., 1981. *The discovery of New Mexico by the Franciscan monk fray Marcos de Niza in 1539*.

Banks, J.B., 1962. *The Endeavour Journal of Joseph Banks 1768–71*. Ed. by J. C. Beaglehole.

Barber, R.T., 1988. 'The ocean basin ecosystem', in Alberts, J.J. and L.R. Pomeroy (eds), *Concepts of ecosystem ecology*.

Barber, R.T., 1990. 'Nutrients and productivity during the 1982–3 El Niño', in Glynn, P.W. (ed), *Global ecological consequences of the 1982–83 El Niño-Southern Oscillation*.

Barber, R.T. and F. P. Chavez, 1983. 'Biological consequences of El Niño.' Science 222: 1203–10.

Barber, R.T. and F. P. Chavez, 1986. 'Ocean variability in relation to living resources during the 1982–83 El Niño.' Nature 319: 279–285.

Barbosa, D., 1918. *Book of Duarte Barbosa; an account of the countries bordering on the Indian Ocean*. Trans. by M. Dames.

Barry, J.M., 1997. *Rising Tide; the great Mississippi flood of 1927 and how it changed America*.

Barth, Dr H., 1965: *Travels and Discoveries in North and Central Africa 1849–1855*.

Basher, R.E. and X. Zheng, 1995. 'Tropical cyclones in the southwest Pacific: spatial patterns and relationships to Southern Oscillation and sea surface temperature.' J. Clim. 8(5): 1249–60.

Baylis, M, P.S. Mellor and R. Melswinkel, 1999. 'Horse sickness and ENSO in South Africa.' Nature 397:574.

Beatson, A., 1816. 'Tracts relative to the Island of St. Helena; written during a residence of five years.'

Beechey, F.W., 1831. *Narrative of a voyage to the Pacific and Bering Strait*, Vol 1.

Belden, J., 1949. *China shakes the world.*

Bell, B., 1971. 'The dark ages in ancient history: I, The First Dark Age.' Egypt. Am. J. Arch. 75: 1–26.

Bell, B., 1975. 'Climate and the history of Egypt: the Middle Kingdom.' A.J.A. 79: 223–69.

Bennett, J.A., 1976. 'Immigration, blackbirding, labour recruiting? The Hawaiian experience 1877–1887.' J. Pac. Hist. 11: 3–27.

Berger, L. et al, 1998. 'Chytridiomycosis causes amphibian mortality associated with population declines in the rain forests of Australia and Central America.' Proc. Natl. Acad. Sci. USA 95: 9031–6

Bhatia, B., 1963. *Famines in India 1860–1965.*

Binder, P., 1977. *Treasure Islands: the trials of the Ocean Islanders.*

Binford, M.W. et al, 1997. 'Climate variation and the rise and fall of an Andean civilization.' Quat. Res. 47: 235–48.

Birch, J.R. and J.R. Marko, 1986. 'The severity of the iceberg season on the Grand Banks of Newfoundland: an El Niño connection?' Trop. Ocean/Atmosphere News. 36 18–22.

Birkeland, C., 1989. 'The faustian traits of the crown-of-thorns starfish.' Am. Sci. 77: 154–163.

Birmingham, D., 1993. *A concise history of Portugal.*

Blainey, G., 1983. *A land half won.*

Blainey, G., 1988. *The great seesaw: a new view of the world 1750–2000.*

Blaxland, G., 1913. *A journal of a tour of discovery across the Blue Mountains, N.S.W., in the year 1813.* Ed. by F. Walker.

Bloom, K.J., 1993. *The Mississippi Valley's great yellow fever epidemic of 1878.*

Bock, C.A., 1881. *The head-hunters of Borneo: a narrative of travel up the Mahakkam and down the Barito.*

Bohr, P.R., 1972. *Famine in China and the missionary: Timothy Richard as relief administrator and advocate of national reform, 1876–1884.*

Bouma, M.J. and C. Dye, 1997. 'Cycles of malaria associated with El Niño in Venezuela'. J. Am. Med. Ass. 278 (21): 1772–4.

Bouma, M.J. and H.J. van der Kaay, 1996. 'The El Niño Southern Oscillation and the historic malaria epidemics on the Indian subcontinent and Sri Lanka: an early warning system for future epidemics?' Trop. Med. Int. Health 1 (1): 86–96.

Boua, M.J., R.S. Kovats, J. St.H. Cox and A. Haines, 1997. 'Global assessment of El Niño's disaster burden.' Lancet 350: 1435–8.

Bourget, S. (in press). 'Rituals of Sacrifice: Its practice at Huaca de la Luna and Its Representation in Moche iconography.' In Pillsbury, J. (ed), *Moche Art and Archaeology*

Bourget, S. (in press). 'Children and Ancestors: Ritual Practices at the Moche Site of Huaca de la Luna, North Coast of Peru'. In E.P. Benson, and A. Cook (eds), *Ritual Sacrifice in Ancient Peru: New Discoveries and Interpretations.*

Bourke, A. and H. Lamb, 1993. *The spread of the potato blight in 1845–6 and the accompanying wind and weather patterns.*

Bowman, I., 1931. *The pioneer fringe.*

Bridgman, H.A., 1983. 'Could climatic change have had an influence on the Polynesian migrations?' Palaeogeog, Palaeoclim, Palaeoecol. 41: 193–206.

Brierley, A. and K. Reid, 1999. 'Kingdom of the krill.' New Scientist 162 (2182): 37–41.

Brookfield, H. and B. Allen, 1989. 'High-altitude occupation and environment,' Mountain Res. and Development 9(3): 210–223.

Brookfield, H., L. Potter and Y. Bryon, 1995. *In place of the forest: environmental and socio-economic transformation in Borneo and the Eastern Malay peninsula.*

Brooks, R.H., 1971. 'Human Response to recurrent drought in northeast Brazil.' Prof'. Geog. 23(1): pp. 40–4.

Brown, R., 1985. *Voyage of the iceberg.*

Brundage, B.C., 1979. *The Fifth Sun.*

Bryant, A.W.M., 1957. *The turn of the tide, 1939–1943.*

Bryant, E.B., 1991. *Natural Hazards.*

Buddemeier, R.W. and D.G. Fautin, 1993. 'Coral bleaching as an adaptive mechanism: a testable hypothesis.' Bioscience 43(5): 320–6.

Buddemeier, R.W., 1997. 'Making light work of adaptation.' Nature 388: 229.

Bunkley-Williams, L. and E.H. Williams Jr., 1990. 'Global assault on coral reefs.' Nat. Hist. 4/90: 46–54.

Buttinger, J. 1967. *Vietnam: A dragon embattled.*

Butzer, K.W., 1976. *Early hydraulic civilization in Egypt: a study in cultural ecology.*

Cahill, R.A., 1990. *Disasters at sea, Titanic to Exxon Valdez.*

Caldecott, J. and S. Caldecott, 1985. 'A horde of pork.' New Sc. 33–5.

Cane, M.A., G. Eshel and R.W. Buckland, 1994. 'Forecasting Zimbabwean maize yield using eastern equatorial Pacific sea surface temperature.' Nature 370: 204–5.

Cannon, M., 1987. *The exploration of Australia.*

Caputi, N. et al, 1996. 'Effect of the Leeuwin current on the recruitment of fish and invertebrates along the West Australian coast.' Mar. Freshwater Res. 47: 147–55.

Cashdan, E.A., 1986. 'Coping with risk: reciprocity among the Basarwa of northern Botswana,' Man 20: 454–74.

Cashdan, E.A., 1980. 'Egalitarianism among hunters and gatherers.' Am. Anthr. 82 (1): 116–120.

Cashdan, E.A., 1989. 'Hunters and gatherers: economic behaviour in bands', in Plattner, S. (ed.), *Economic anthropology.*

Caviedes, C.N., 1984. 'El Niño 1982–83.' Geog. Rev. 74 (3): 267–90.

Caviedes, C.N., 1991. 'Five hundred years of hurricanes in the Caribbean: their relationship with global climatic variabilities.' Geo. J. 23 (4): 301–10.

Caviedes, C.N. and P.R. Waylen, 1993. 'Anomalous westerly winds during El Niño events: the discovery and colonization of Easter Island.' Applied Geog. 13: 123–34.

Chan, A., 1982. *Glory and fall of the Ming dynasty.*

Chang, W.Y.B. and G. King, 1994. 'Centennial climate changes and their global associations in the Yangtze River Delta, China and subtropical Asia.' Clim. Res. 4: 95–103.

Chang, W.Y.B., 1997. 'ENSO: Extreme climate events and their impacts on Asian deltas.' J. Am. Water Res. Ass. 33 (3): 605–14.

Changnon, S.A. (ed), 1996. *The great flood of 1993: causes, impacts and responses.*

Chapman, T. et al, 1982. *Niue – A history of the island.*

Chastel, O., H. Weimerskirch and P. Jouventin, 1992. *Annual variability in reproductive success and survival of an Antarctic seabird, the snow petrel: a 27 year study.*

Chatwin, B., 1987. *The Songlines.*

Christophers, S.R., 1911. *Malaria in the Punjab.*

Cieza de Leon, P. (ed.), 1959. *The Incas*, trans. by H. de Onis.

Clark, C.M.H., 1981. *History of Australia: Vol. 5, the people make the laws 1888–1915.*

Claxton, R.H., 1986. *Weather based hazards in colonial America.*

Claxton, R.H. (ed). *Investigating natural hazards in Latin American history.*

Claxton, R.H., 1993. 'The record of drought and its impact in colonial Spanish America.' In Herr, R. (ed) *Themes in rural history of the western world.*

Coetzee, D., 1974. 'The Niger coup: why drought and uranium don't mix.' African Development 8 (6): 16–22.

Cohen, J.M., 1981: *The discovery and conquest of Peru.*

Cohen, P.A., 1997 *History in three keys: the Boxers as event, experience and myth.*

Colson, E., 1979. 'In good years and bad: food strategies of self-reliant societies.' J. Anthrop. Res. 35: 18–29.

Connell, J., 1986. *New Caledonia or Kanaky?: The political history of a French colony.*

Conrad, R.E., 1993. *The destruction of Brazilian slavery, 1850–1888.*

Cook, J., 1777. *A voyage toward the south pole and round the world.*

Cook, J., 1785. *A voyage to the Pacific Ocean in the years 1776–79 and 1780.*

Cook, J., 1793. *Captain Cook's journal during his first voyage round the world.* Ed. by J. Wharton.

Costa, D.P. and T.M. Williams, 1999. 'Marine mammal energetics'. In Reynolds, J.E. and S.A. Rommel (eds), *Biology of marine mammals.*

Crespi, Fray J., 1971. *Missionary explorer on the Pacific coast, 1769–1774,* edited by H.E. Bolton.

Cronwright-Schreiner, S.C., 1925. *The migratory springbucks of South Africa.*

Cunha, E. Da, 1943. *Rebellion in the backlands.* Trans. by S. Putnam.

Croxali, J.P., K. Reid and P.A. Prince, 1999. 'Diet, provisioning and productivity reponses of predators to differences in availability of Antarctic krill.' Mar. Ecol. Prog. Ser. 177: 115–31.

Curcio, L.A., 1997. 'Virgin of Los Remedios.' In Werner M.S. (ed), *Encyclopedia of Mexico.*

Darwin, C., 1933: *Diary of the voyage of the HMS Beagle,* ed. by N. Barlow.

Darwin, C., 1859. *The origin of species.*

Davidson, O.G., 1998. *The enchanted braid: coming to terms with nature on the coral reef.*

Davie, N., 1987. *The Titanic.*

Davies, N., 1977. *The Aztecs; a history.*

Day, A.G., 1940. *Coronado's quest.*

Dellenbaugh, F.S., 1905. *Breaking the Wilderness: the story of the conquest of the Far West.*

Dening, G., 1980. *Islands and beaches, a discourse on a silent island.*

Desender, K., L. Baert and J-p. Maelfait, 1992. 'El Niño events and establishment of ground betles in the Galápagos archipelago.' Bull. Inst. Royal Sc. Ento. Belg. 62: 67–74.

DeVries, T.J. et al, 1997. 'Determining the early history of El Niño.' Science 276: 965–7.

Diamond, J.M., 1985. 'Mass mortality in Pacific seabirds.' Nature 315: 712–13.

Dias, J.R., 1981. 'Famine and disease in the history of Angola, c. 1830–1930.' J. Afr. Hist. 22: 349–78

Diaz, H. F. and G.N. Kiladis, 1995. 'Climatic variability on decadal to century time-scales.' In Henderson-Sellers, A. (ed), *Future climates of the world: a modelling perspective.*

Diaz, H.F. and G.J. McCabe, 1999. 'A possible connection between the 1878 yellow fever epidemic in the southern United States and the 1877–78 El Niño episode.' Bull. Am. Met. Soc. 80 (1): 21–27.

Dickman, C. R., M. Predavec and F. J. Downey, 1995. 'Long-range movements of small mammals in arid Australia: implications for land management.' J. Arid Envts 31: 441–452.

Dilley, M., and B.N. Heyman, 1995. 'ENSO and disaster: droughts, floods and El Niño/Southern Oscillation warm events.' Disasters 19 (3): 181–93.

Dillon, M.O. and P.W. Rundel, 1989. 'The botanical response of the Atacama and Peruvian desert floras to the 1982–83 El Niño event.' In Glynn, P. (ed), *Global ecological consequences of the 1982–83 El Niño-Southern Oscillation.*

Donnan, C.B., 1990. 'An assessment of the valdity of the Naymlap dynasty.' In Moseley, M.E. and A. Cordy-Collins (eds), *The northern dynasties, kingship and statecraft in Chimor.*

Douglas, B., 1998. *Across the great divide: journeys in history and anthropology.*

Douglas, B., 1970. 'A contact history of the Balad people of New Caledonia, 1774–1845.' Poly. Soc. J. 79 (2): 180–200.

Douglas, M.S., 1958. *Hurricane.*

Dove, M.R. and D.M. Kammen, 1997. 'The epistemology of sustainable resource use: managing forest products, swiddens, and high-yielding variety crops.' Human Org. 56 (1): 91–101.

Dove, M.R., 1993. 'The responses of Dayak and bearded pig to mast-fruiting in Kalimantan: an analysis of nature-culture analogies.' In Hladik, C.M. et al (eds), *Tropical forests, people and food: biocultural interactions and applications to development.*

DuBois, V.D., 1974. 'The drought in Niger.' Am. Uni. Field Staff, West Africa Series XV, No 4–6.

Duffy D.C. et al, 1988. 'Comparison of the effects of El Niño and Southern Oscillation on birds in Peru and the Atlantic Ocean.' In Ouellet, H. (ed), Acta XIX Congressus Int. Ornith. Vol. 2.

Duffy, D.C., 1989. 'Seabirds and the 1982–1984 El Niño-Southern Oscillation.' In Glynn, P. (ed), *Global ecological consequences of the 1982–83 El Niño-Southern Oscillation.*

Duiker, W.J., 1996. *The communist road to power in Vietnam.*

Dunbar, R.B., G.M. Wellington, M.W. Colgan and P.W. Glynn, 1994, 'Eastern Pacific sea surface temperature since 1600AD.' Paleoceanography 9(2); 291–315.

Dunstan, H., 1975. 'The late-Ming epidemics: A preliminary survey.' Ching-shih wen-ti 3 (3) 1–59.

Duran, D., 1964. *The Aztecs: the history of the Indies of New Spain.* Trans. by D. Heyden and F. Horcasitas.

Eakin, M.E., 1988. 'Avoidance of damselfish lawns by the sea-urchin *Diadema mexicanum* at Uva Island, Panama.' Proc. 6th Int. Coral Reef Symp. Vol. 2.

Eakin, M.E., 1992. Post-El Niño 'Panamanian reefs: less accretion, more erosion and damselfish protection.' Proc. 7th. Int. Coral Reef Symp. Vol. 1.

Eakin, M.E., 1996. 'Where have all the carbonates gone? A model comparison of calcium carbonate budgets before and after the 1982–1983 El Niño at Uva Island in the Eastern Pacific.' Coral Reefs 15: 109–119.

Eberhard, W., 1977. *A history of China.*

Editors of Encyclopaedia Britannica, 1978. *Disaster! When nature strikes back.*

Eldredge, E.A., 1992. 'Sources of Conflict in Southern Africa, c1800–30: The Mfecane Reconsidered.' J. of Afr. Hist. 33 (1) 1–35.

Elvin, M., 1998. 'Who was responsible for the weather? moral meteorology in late-imperial China.' OSIRIS, 13: 213–237.

Endfield, G.H. and S.L. O'Hara, 1997. 'Conflicts over water in "The Little Drought Age" in Central Mexico.' Envt. and History 3: 255–72.

Enfield, D.B., 1989. 'El Niño, past and present.' Rev. Geophys. 27 (1): 159–87.

Enfield, D., 1988. 'Is El Niño becoming more common?' Oceano-
    graphy Mag. 1: 23–27.
Epstein, P.R., 1997. 'Climate, ecology, and human health.' Con-
    sequences 3 (2): 3–19.
Eriksen, C.C., 1993. 'Equatorial ocean response to rapidly trans-
    lating wind bursts.' J. Phys. Oceanography 23: 1208–1230.
Estrada, E. and B. Meggers, 1961. 'A complex of traits of probable
    transpacific origin on the coast of Ecuador.' Am. Anthr. 63: 913–
    39.
Evans, E.W.P., 1945. *Timothy Richard: a narrative of Christian
    enterprise and statesmanship in China.*
Fagan, B.M. (ed), 1996. *The Oxford companion to archaeology.*
Feng, Y-L., 1976. *A short history of Chinese philosophy.*
Ferdon, E.N., 1993. *Early observations of Marquesan culture,
    1595–1813.*
Fernando, R., 1980. *Famine in Cirebon residency in Java, 1844–
    1850; a new perspective on the cultivation system.*
Figes, O., 1996. *A people's tragedy: the Russian revolution, 1891–
    1924.*
Finney, B.R., 1985. 'Anomalous westerlies, El Niño, and the
    colonization of Polynesia.' Am. Anthrop. 87: 9–26.
Finney, B., 1994. *Voyage of discovery: a cultural odyssey through
    Polynesia.*
Fisher, H.H., 1927. *The famine in Soviet Russia 1919–23.*
Fitzgerald, C.P., 1986. *China: a short cultural history.*
Flannery, T., 1994. *The future eaters.*
Forbes, H.O., 1914. 'Notes on Molina's pelican.' Ibis 2: 403–20.
Forrester, G. and R.J. May, 1998. *The fall of Soeharto.*
Forster, G., 1777. *A voyage round the world in his Britannic
    Majesty's sloop Resolution during the years 1772, 3, 4, and 5.*
Fosberg, F.R. (ed), 1970. *Man's place in the island ecosystem.*
Fraedrich, K., 1994. 'An ENSO impact on Europe?: A review.'
    Tellus 46A: 541–52.
Frampton, S., J. Chaffey, A. McNaught and J. Hardwick, 1996.
    Natural Hazards causes, consequences and management.
Gardiner, R. and D. van der Vat, 1995. *The riddle of the Titanic.*

Ghose, H.-P., 1943. *The famine of 1770.*

Gibbs, H.L. and P.R. Grant, 1987. 'Ecological consequences of an exceptionally strong El Niño event on Darwin's finches.' Ecology 68 (6): 1735–46.

Gibbs, H.L. and P.R. Grant, 1987. 'Oscillating selection on Darwin's finches.' Nature 327: 511–13.

Gibbs, W.J., 1998. 'The origins of Australian meteorology.' Aust. Bureau of Met. Papers No. 12.

Gibson, C., 1964. *The Aztecs under Spanish rule; a history of the Indians of the Valley of Mexico.*

Gilbert, A., 1981. 'The state and nature in Australia.' Aust. Cult. Hist. 1: 9–28.

Gilliomee, H., 1979. 'The Eastern frontier, 1770–1812.' In Elphick, R. and H. Gilliomee (eds), *The shaping of South African society, 1652–1820.*

Gill, C.A., 1921. 'The Role of Meteorology in Malaria.' Ind. J. Med. Res. 8: 633–93.

Gill, Rev. W., 1855. *Gems from the Coral Islands; or incidents of contrast between savage and Christian life in the South Sea islanders.*

Glantz, M.H. (ed), 1994. *Drought follows the plough.*

Glantz, M.H., 1996. *Currents of change: El Niño's Impact on climate and society.*

Glantz, M.H., R.W. Katz and N. Nicholls, 1991. *Teleconnections linking worldwide climate anomalies.*

Glynn, P., 1988. 'El Niño warming, coral mortality and reef framework destruction by echinoid bioerosion in the eastern Pacific.' Galaea 7: 129–60.

Glynn, P.W. and W.H. de Weerdt, 1991. 'Elimination of two reef-building hydrocorals following the 1982–3 El Niño warming event.' Science 235 (5015): 69–71.

Goodrich, L.C. (ed), 1976. *Dictionary of Ming biography, 1368–1644.* Pt 1.

Grant, B.R. and P.R. Grant, 1993. 'Evolution of Darwin's finches caused by a rare climatic event.' Proc. R. Soc. Lond. B 251:111–17.

Grant, P.R. and B.R. Grant, 1987. 'The extraordinary El Niño event of 1982–3: effects on Darwin's finches on Isla Genovesa, Galapagos.' Oikos 49: 55–66.

Green, C.S., 1955. *Karoo.*

Green, P.M., D. Legler, C.J. Miranda, and J.J. O'Brien, 1997. 'The North American climate patterns associated with the El Niño-Southern Oscillation.' COAPS Proj. Rep. Ser. 97–1.

Greenfield, G.M., 1992. 'The great drought and elite discourse in imperial Brazil.' Hisp. Am Hist. Rev. 72 (3): 375–400.

Gribbin, J. and M. Gribbin, 1997. *Watching the weather.*

Grove, R.H., 1997. *Ecology, climate and empire: colonialism and global environmental history 1400–1940.*

Guerra, C.G. and G.K. Portflitt, 1991. 'El Niño effects on pinnipeds in Northern Chile.' In Trillmich, F. and K.A. Ono (eds), *Pinnipeds and El Niño: responses to environmental stress.*

Guillory, D., 1996. *When the waters recede; rescue and recovery during the Great Flood.*

Guzman, H.M. and J. Cortés, 1992. 'Cocos Island coral reefs after the 1982–83 El Niño disturbance.' Rev. Biol. Trop. 40 (3): 309–24.

Hales, S.P. Weinstein and A. Woodward, 1996. 'Dengue fever epidemics in the South Pacific: driven by El Niño Southern Oscillation?' Lancet 348: 1664–5.

Hall, A.L., 1978. *Drought and irrigation in Northeast Brazil.*

Hallenbeck, C., 1949. *The journey of Fray Marcos de Niza.*

Hamilton, K. and R.R. Garcia, 1986. 'El Niño/Southern Oscillation events and their associated midlatitude teleconnections 1531–1841.' Bull. Met. Soc. 67 (11): 1354–60.

Hanna, W.L., 1979. *Lost harbor: the controversy over Drake's Californian anchorage.*

Harms, R.W., 1987. *Games against nature; an eco-cultural history of the Nunu of equatorial Africa.*

Hart, J.M., 1989. *Revolutionary Mexico: the coming and process of the Mexican Revolution.*

Hassan, F.A., 1994. 'Population ecology and civilization in ancient Egypt.' In Crumley, C.L. (ed), *Historical ecology, cultural knowledge and changing landscapes.*

Hassig, R., 1981. 'The famine of One Rabbit: ecological causes and social consequences of a pre-Colombian calamity.' J. Anthrop. Res. 37: 171– 81.

Hatch, S., 1987. 'Did the 1982–1983 El Niño-Southern Oscillation affect seabirds in Alaska?' Wilson Bull. 99 (3): 468–74.

Hayden, B., 1992a. 'Ecology and complex hunter/gatherers.' In Hayden, B. (ed) *A complex culture of the British Colombia Plateau.*

Hayden, B., 1992b. 'Ecology and culture.' In Hayden, B. (ed), *A complex culture of the British Colombia Plateau.*

Hayden, B., 1993. *Archaeology: the science of once and future things.*

Hayden, B., 1975. 'The carrying capacity dilemma: an alternative approach.' *Am. Antiq. 40 (2pt.2) 11–21.*

Heathcote, R.L., 1988. 'Drought in Australia: still a problem of perception?' GeoJournal 16 (4): 387–97.

Heathcote, R.L., 1969, 'Drought in Australia: a problem in perception.' Geog. Rev. 59: 175–94.

Heathcote, R.L., 1987. 'Images of a desert? Perception of arid Australia.' Aust. Geog. Studies 25: 325.

Hemming, J., 1983. *The conquest of the Incas.*

Herring, R.S., 1979. 'Hydrology and chronology: the Rodah Nilometer as an aid to dating interlacustrine history.' In Webster, J.B. (ed), *Chronology, migration and drought in interlacustrine Africa.*

Heyerdahl, T., 1950. 'The voyage of the raft Kon-Tiki.' Geog J. 115: 20–41.

Heyerdahl, T., D. Sandweiss and A. Narvaez, 1995. *Pyramids of Tucume: the quest for Peru's forgotten city.*

Heywood, R.B., I. Everson and J. Priddle, 1985. 'The absence of krill from the South Georgia zone, winter 1983.' Deep-Sea Res. 32 (3): 369–78.

Hitchcock, R.K., 1979. 'The traditional response to drought in Botswana.' In Hinchey, M.T. (ed), Proc. Symp. on Drought in Botswana.

Hittell, T.H., 1885. *History of California.*

Hodell, D.A., J.H. Curtis and M. Brenner, 1995. 'Possible role of climate in the collapse of Classic Maya civilization.' Nature 375: 391–4.

Holt and Seaman, 1976. 'The scope of the drought.' In Hussein, A.M. (ed), *Rehab: drought and famine in Ethiopia*.

Holt, P.M. and M.W. Daly, 1988. *A history of the Sudan; from the coming of Islam to the present day*.

Hosie, A., 1878. 'Droughts in China AD 620 to 1643.' J. North-China Branch Roy. Asiatic Soc. 12: 51–89.

Hoyt, W.G. and W.B. Langbein, 1955. *Floods*.

Hugenholtz, R., 1986. 'Famine and food supply in Java 1830–1914.' In Bayly, C.A. and D. Kilft (eds), *Two colonial empires*.

Hughen, K.A., D.P. Schrag amd S.B. Jacobsen, 1999. 'El Niño during the last interglacial period recorded by a fossil coral from Indonesia.' Geophys. Res. Letters 26 (20): 3129–32.

Human Rights Watch, 1991. *Evil Days, thirty years of war and famine in Ethiopia*.

Humboldt, A. de and A. Bonpland, 1819. *Personal narrative of travels to the equinoctial regions of the new continent during the years 1799–1804*.

Hunter, W.W., 1897. *Annals of rural Bengal*.

Idyll, C.P., 1973. 'The anchovy crisis.' Sc. Am. 228: 22–29.

Iliffe, J., 1990. *Famine in Zimbabwe, 1890–1960*.

Iliffe, J., 1995. *Africans: the history of a continent*.

Inoue, T. et al, 1993. 'Population dynamics of animals in unpredictably-changing tropical environments.' J. Biosc. 18 (4): 425–55.

Irwin, G., 1992. *The prehistoric exploration and colonisation of the Pacific*.

Jeal, T., 1973. *Livingstone*.

Jenkins, J., 1975. *Diary of a Welsh Swagman, 1869–1894*.

Johnston, G., 1964. *My Brother Jack*.

Joy, W., 1964. *The explorers*.

Joyner, T., 1992. *Magellan*.

Junker, W., 1890. *Travels in Africa during the years 1875–1878*.

Kaplan, A. et al, 1998. 'Analyses of global sea surface temperature 1856–1991.' J. Geophys. Res. 103 (C9): 18567–89.

Kapuściński, R., 1983. *The emperor, downfall of an autocrat.*

Katz, F., 1972. *The ancient American civilisations.*

Keast, A., 1981. 'The evolutionary biogeography of Australian birds.' In Keast, A. (ed), *Ecological biogeography of Australia.*

Keeble, J., 1994. *Out of the channel: the Exxon Valdez oil spill in Prince William Sound.*

Kerr, R.A., 1998. 'Models win big in forecasting El Niño.' Science 280: 522–3.

Kiladis, G.N. and H.F. Diaz, 1989. 'Global climatic anomalies associated with extremes in the Southern Oscillation.' J. Clim. 2 (9): 1069–90.

Kiladis, G.N. and H.F. Diaz, 1986. 'An analysis of the 1877–78 ENSO episode and comparison with 1982–3.' Mon. Weather Rev. 114: 1035–47.

Kimambo, I.N., 1969. *A political history of the Pare of Tanzania, c1500–1900.*

Kimberly, W.B., 1897. *History of West Australia.*

Knutson, T.R. and S. Manabe, 1994. 'Impact of increased CO^2 on simulated ENSO-like phenomena.' Geophys. Res. Letters 21 (21): 2295–98.

Kolata, A., 1996. *Valley of the spirits: a journey into the lost realm of the Aymara.*

Konnen, G.P., P. Jones, M. Kaltofen and R. Allan, 1998. 'Pre–1866 extensions of the Southern Oscillation Index using early Indonesian and Tahitian meteorological readings.' J. Climate 11 (9): 2325–39.

Kovar, A. 1970. 'The physical and biological environment of the Basin of Mexico.' In Sanders, W.T.A. Kovar, T. Charlton and R.A. Diehl, *The Teotihuacan Valley project: the natural environment, contemporary occupation and 16th Century population of the valley.*

Kovats, R.S., M.J. Bouma and A. Haines, 1999. *El Niño and health. WHO report.*

La Barre, W., 1948. 'The Aymara Indians of the Lake Titicaca Plateau, Bolivia'. Memoirs Am. Anthrop. Assoc. No. 68.

La Billardiére, J.J.H de., 1800. *An account of a voyage in search of La Perouse in the years 1791, 1792 and 1794.*

Laband, J. and I. Knight, 1996. *The Anglo-Zulu war*.

Laband, J., 1997. *The rise and fall of the Zulu nation*.

LaFrankie, J.V. and H.T. Chan, 1991. 'Confirmation of sequential flowering in Shorea (Dipterocarpaceae).' Biotropica 23 (2): 200–3.

Lamb, H. H., 1995. *Climate, history and the modern world*.

Langdon, R., 1989. 'When the blue-egg chickens come home to roost.' J. Pac. Hist. 24: 164–90'.

Langdon, R., 1993. 'The banana as a key to early American and Polynesian history.' J. Pac. Hist. 28(1): 15–36.

Latif, M. et al, 1994. 'A review of ENSO prediction studies.' Clim. Dynamics 9: 167–79.

Latz, P., 1995. *Bushfires and bushtucker: aboriginal plant use in Central Australia*.

Laurie, A., 1989. 'Effects of the 1982–83 El Niño-Southern Oscillation event on marine iguana populations on Galapagos.' In Glynn, P. (ed), *Global ecological consequences of the 1982–83 El Niño-Southern Oscillation*.

Laws, E.A., 1997. *El Niño and the Peruvian anchovy fishery*.

Laws, R.M., 1985. 'The ecology of the Southern Ocean.' Am. Sci. 73: 26–40.

Lawson, H., 1979. *Poetical works of Henry Lawson*.

LeComte, D., 1989. 'The rains return to the tropics.' Weatherwise 42 (1): 8–12.

Lee, R.B., 1968. 'What hunters do for a living, or, how to make out on scarce resources.' In Lee, R.B. and I. DeVore (eds), *Man the hunter*.

Lee, R.B., 1976. '!Kung spatial organization: an ecological and historical perspective.' In Lee, R.B. and I. DeVore (eds), *Kalahari hunter-gatherers: studies of the !Kung San and their neighbours*.

Legum, C., 1975. *The fall of Haile Selassie's empire*.

Lehodey, P. et al, 1997. 'El Niño Southern Oscillation and tuna in the western Pacific.' Nature 389: 715–17.

Lenanton, R.C. et al, 1991. 'The influence of the Leeuwin current on coastal fisheries of Western Australia.' J. Royal Soc. West Aust. 74: 101–14.

Levinson, M., R G. Ward and J. W. Ward, 1972. 'The settlement of Polynesia: a report on a computer simulation.' Arch. Phys. Anthrop. in Oceania 7 (3): 234–45.

Lewins, R. et al, 1994. 'The emergence of new diseases.' Am. Sc. 82: 52–60.

Li, T. and A. Reid, 1993. *Southern Vietnam under the Nguyen.*

Lichtheim, M., 1973. *Ancient Egyptian Literature.*

Lightoller, C., 1935. *Titanic and other ships.*

Limpus, C., 1989. 'Environmental regulation of green turtle breeding in Eastern Australia.' Proc. 9th Annual Workshop on Sea Turtle Cons. and Biol.

Limpus, C.J. and N. Nicholls, 1988. 'The Southern Oscillation regulates the annual numbers of green turtles (*Chelonia mydas*) breeding around Northern Australia.' Aust. J. Wildl. Res. 15: 157–61.

Lindesay, J.A. and C.H. Vogel, 1990. 'Historical evidence for Southern Oscillation – Southern African rainfall relationships.' Int. J. Climatol. 10: 679–89.

Lindsay, D. and L.A. Wells, 1893. *Journal of the Elder scientific exploring expedition, 1891–1892.*

Livingstone, D., 1960. *Livingstone's private journal 1851–1853.*

Livingstone, D., 1961. *Livingstone's missionary correspondence, 1841–1856.* Ed. by I. Schapera.

Llosa, M.V., 1986. *The war at the end of the world.*

Loeb, E.M., 1971. *History and traditions of Niue.*

Loewe, M. and E.L. Shaughnessy, 1999. *The Cambridge history of China.*

Love, G. 1987. 'Banana prawns and the Southern Oscillation Index.' Aust. Met. Mag. 35: 47–9.

Lovegrove, B.G., 1996. 'The low basal metabolic rates of marsupials: the influence of torpor and zoogeography.' In Geiser, F., A.J. Hulbert and S.C. Nicol (eds), *Adaptations to the cold.* 10th Int. Hibernation Symp.

Lovegrove, B.G., 1993. *The living deserts of southern Africa.*

Lovejoy, P.E. and S. Baier, 1975. 'The desert-side economy of the central Sudan.' Int. J. Afr. Hist.Stud. 8(4): 551–81.

Loveman, B., 1988. *Chile*.

Low, B.S., 1978. 'Environmental uncertainty and the parental strategies of marsupials and placentals.' Am. Nat. 112 (No.983): 197–213.

Lucas, J., 1998. *War on the eastern front: the German soldier in Russia, 1941–45*.

Luling, V., 1975. 'Drought in Somali literature.' In Lewis, I. (ed), *Abaar: The Somali drought*.

Lutz, C.H., 1994. *Santiago de Guatemala, 1541–1773: city, caste and colonial experience*.

Lyons, M., 1986. *The totem and the tricolour: a short history of New Caledonia since 1774*.

Macaulay, T.B., 1852. *Lord Clive*.

Mallory, W.H., 1926. *China: land of famine*.

Marcos, J.G., 1977. 'Cruising to Acapulco and back with the thorny oyster set: a model for a lineal exchange system.' J. Steward Anthr. Soc. 9 (1–2): 99–132.

Marcos, J.G. 1998. 'The Manteño-Huancavilca merchant lords of Ancient Ecuador; their predecessors and their trading partners.' Conf. Pre-Colombian Soc. of Wash.

Marcus, H.G., 1994. *A history of Ethiopia*.

Marshall, E., 1993. 'Hantavirus outbreak yields to PCR.' Science 262: 832–6.

Maylam, P., 1986. *A history of the African people of South Africa, from the early Iron Age to 1970's*.

Macculloch, J.R., 1841. *A dictionary, geographical, statistical, and historical, of the various countries, places, and principal natural objects of the world*.

Macdonald, B., 1996. 'Lawrence Wells'. In Macdonald, B. and A. Rymill, *The Explorers*.

McDonald, B., 1982. *Cinderellas of the empire*.

Macgregor, G., 1937. *Ethnology of Tokelau Island*.

McKinlay, J., 1863. *Tracks of McKinlay and party across Australia*.

McLynn, F.J., 1997. *Napoleon: a biography*.

Meehl, G.A. and W.M. Washington, 1996. 'El Niño-like climate change in a model with increased atmospheric CO^2 concentrations.' Nature 382: 56–60.

Meek, C.K., 1931. *A Sudanese kingdom.*

Meggers, B., 1995. 'Amazonia on the eve of European contact: ethnohistorical, ecological, and anthropological perspectives.' J. Am. Arch. 8: 91–115.

Meggers, B., 1994. 'Archaeological evidence for the impact of mega-niño events on Amazonia during the past two millenia.' Clim. Change 28: 321–338.

Meinen, C.S. and M.J. McPhaden, 2000. 'Observations of warm water changes in the equatorial Pacific and their relationship to El Nino and La Nina.' J. Clim. (In press).

Merlen, G., 1985, in Robinson, G. and E.M.del Pino (eds). *El Niño en las Islas Galapágos: el evento de 1982–1983.*

Merlen, G., 1984. 'The 1982–83 El Niño: some of its consequences for Galapágos wildlife.' Oryx 18 (4): 210–4.

Meyer, M.C. and W.L. Sherman, 1979. *The course of Mexican history.*

Meyer, M.C., 1984. *Water in the Hispanic southwest, a social and legal history, 1550–1850.*

Meyerson, E.A. et al, 2000. 'The polar expression of ENSO and sea ice variability as recorded in a South Pole ice core.' (In press).

Miller, J.C., 1982. 'The significance of drought, disease and famine in the agriculturally marginal zones of west-central Africa.' J. Afr. Hist. 23: 17–61.

Miller, J.C., 1988. *Way of death: merchant capitalism and the Angolan slave trade, 1730–1830.*

Milne, A., 1986. *Floodshock; the drowing of planet earth.*

Milton, J., 1983. *The edges of war: a history of the Frontier Wars.*

Minc, L.D., 1986. 'Scarcity and survival: the role of oral tradition in mediating subsistence crises.' J. Anthrop. Arch. 5: 39–113.

Mitchell, T.L., 1838. *Three expeditions into the interior of Australia.*

Monk, K., Y. de Fretes and G. Reksodiharjo-Lilley, 1997. *The ecology of Nusa Tenggara and Maluku.*

Moors, H.J., c.1880. *The Tokanoa: a plain tale of some strange adventures in the Gilberts*, compiled from the diary of John T. Bradley.

Morison, S.E., 1974. *The European discovery of America: The southern voyages.*

Morris, D.R., 1965. *The washing of the spears.*

Morton, S.R., 1986. 'Land of uncertainty: the Australian arid zone.' In Recher, H. F., D. Lunney, I. Dunn (eds), *A Natural Legacy.*

Moseley, M.E., 1975. *The maritime foundations of Andean civilizations.*

Moseley, M.E., 1999. 'Convergent catastrophe: past patterns and future implications of collateral natural disasters in the Andes.' In Ollver-Smith, A. and S.M. Hoffman (eds), *The angry earth: disaster in anthropological perspective.*

Moseley, M.E, R.A. Feldman, and C.R. Ortloff, 1981. 'Living with crises: human perception of process and time.' In Nitecki, M. (ed), *Biotic crises in ecological and evolutionary time.*

Moseley, M.E., 1992. *The Incas and their ancestors; the archaeology of Peru.*

Moseley, M.E., 1987. 'Punctuated equilibrium: searching the ancient record for El Niño.' The Quarterly Rev. Arch. 8: 7–10.

Moseley, M.E. D. Wagner, and J.B. Richardson III, 1992. 'Space shuttle imagery of recent catastrophic change along the arid Andean coast.' In Johnson, L.L. and M. Stright (eds), *Paleoshorelines and prehistory: an investigation into method.*

Mostert, N., 1992. *Frontiers: the epic of South Africa's creation and the tragedy of the Xhosa people.*

Mote, F.W. and D. Twitchett, 1988. *Cambridge history of China: The Ming dynasty.*

Murphy, R.C., 1926. 'Oceanic and climatic phenomena along the west coast of South America during 1925.' Geog. Rev. 16: 26–54.

Murton, B., 1984. 'Spatial and temporal patterns of famine in Southern India before the famine.' In Currey, B. and G. Hugo (eds), *Famine as a geographical phenomenon.*

Musgrave, S., 1979. *The wayback.*

Myres, M. T., 1985. 'A southward return migration of Painted Lady butterflies, *Vanessa Cardui*, over southern Alberta in the fall of 1983, and biometeorological aspects of their outbreaks into North America and Europe.' Can. Field. Nat 99: 147–55.

Mysak, L., 1986. 'El Niño, international variability and fisheries in the Northeast Pacific Ocean.' Can. J. Fish. Aq. Sc. 43 (2): 464–97.

Nelson, E.C., 1995. *The cause of the calamity: potato blight in Ireland 1845–1847.*

Neumann, J. and H. Flohn, 1987. 'Germany's War on the Soviet Union, 1941–45. Long-range weather forecasts for 1941–2 and climatological studies.' Bull. Am. Met. Soc. Pt 1, 68 (6): 620–30, Pt. 2, 69 (7): 730–5.

Neumann, J., 1992. 'The severe winter 1941–42 in the Eastern Soviet Union: frost casualties among the troops in the Eastern front.' Weather 47 (12): 480–2.

Nials, F. et al, 1979. 'El Niño: the catastrophic flooding of coastal Peru.' Field Mus. Nat. Hist.Bull. 50 (7): 4–14 Pt.1, 50 (8): 4–10 Pt 2.

Nicholls, N., 1988. 'More on Early ENSO's: Evidence from Australian documentary sources.' Bull. Am. Met. Soc 69 (1): 4–6.

Nicholls, N., 1988. 'El Niño-Southern Oscillation and rainfall variability.' J. Climate 1: 418–21.

Nicholls, N., 1989a. 'How old is ENSO?' Clim. Change 14:111–15.

Nicholls, N., 1989b. 'How old is ENSO?' In Donnelly, T.H. and R.J. Wasson (eds), Climanz3: Proc. 3rd Symp. Late Quat. Clim. Hist. Aust.

Nicholls, N., 1993. 'El Niño-Southern Oscillation and vector-borne disease'. Lancet 342: 1284–5.

Nicholls, N. 1997. 'The centennial drought', in Webb, E.K., *Windows on meteorology: Australian perspective.*

Nicholson, S.E., 1999. *Historical and Modern Fluctuations of Lakes Tanganyika and Rukwa and their relationship to rainfall variability.* Clim. Change 41: 53–71.

Norman, F.I. and N. Nicholls, 1991. 'The Southern Oscillation and variations in waterfowl abundance.' Aust. J. Ecol. 16 (4): 485–90.

Normand, C.W.B., 1959. 'Sir Gilbert Walker.' Indian J. Met Geophys. 10: 13–15.

Nunn, G.E., 1934. 'Magellan's route in the Pacific.' Geog. Rev. 24: 615–33.

Oakley, R.O., 1962. *The Southern Sudan, 1883–1898; a struggle for control.*

Obregón, M., 1980. *Argonauts to astronauts.*

Ogilvie, A.E.J., 1995. 'Documentary evidence for changes in the climate of Iceland, AD 1500 to 1800.' In Bradley, R.S. and P.D. Jones (eds), *Climate since AD 1500.*

O'Hara, S.L. and S.E. Metcalfe, 1995. 'Reconstructing the climate of Mexico from historical record's. The Holocene 5 (4): 485–90.

Orlove, B.S. and J.T. Tosteson, 1999. 'The application of seasonal to interannual climate forecasts based on ENSO events: lessons from Australia, Brazil, Ethiopia, Peru and Zimbabwe.' Inst. Int. Studies U. Cal. Berkeley Working Paper 99–3.

Orlove, B.S., J.C.H. Chiang and M.A. Cane, 2000. 'Forecasting Andean rainfall and crop yield from the influence of El Niño on Pleiades visibility.' Nature 403: 68–71.

Ortiz de Montellano, B.R., 1978. 'Aztec cannibalism: an ecological necessity?' Science 200: 611–17.

Osborne, W.S. and M. Davis, 1997. 'Long-term variability in temperature, precipitation and snow cover in the Snowy Mountains: Is there a link with the decline of the southern corroboree frog?' Report for N.S.W. Nat. Parks and Wildl. Serv.

Osborne, W.S., 1989. 'Distribution, relative abundance and conservation status of corroboree frogs.' Aust. Wildl. Res. 16: 537–47.

Oxley, J., 1820. *Journals of two expeditions into the interior of New South Wales.*

Palmer, A, 1997. *Napoleon in Russia.*

Pankhurst, R.P., 1985. *The history of famine and epidemics in Ethiopia: prior to the twentieth century.*

Pankhurst, R., 1967. *The Ethiopian Royal chronicles.*

Parliament, Great Britain, 1990. *Report on the loss of the SS Titanic.*

Parsons, J.B., 1970. *The peasant rebellions of the Late Ming Dynasty.*

Parsonson, G.S., 1972. 'The settlement of Oceania: an examination of the accidental voyage theory.' In J. Golson (ed), *Polynesian navigation.*

Paterson, B., 1993. *The Banjo Paterson Collected Verse.* Ed. by C. Semmler.

Patz, J.A., P.R. Epstein, T.A. Burke and J.M. Balbus, 1996. 'Global climate change and emerging infectious diiseases.' J. Am. Med. Ass. 275 (3): 217–23.

Paulsen, A.C., 1974. 'The thorny oyster and the voice of God: Spondylus and Strombus in Andean prehistory.' Am. Antiq. 39 (4); 597–607.

Pearce, A.F. and B.F. Phillips, 1988. 'ENSO events, the Leeuwin Current, and larval recruitment of the western rock lobster.' J. Cons. Int. Explor. Mer. 45: 13–21.

Pearce, A.F. and B.F. Phillips, 1994. 'Oceanic processes, puerulus settlement and recruitment of the western rock lobster *Panulirus cygnus.*' In Sammarco, P.W. and M.L. Heron (eds), *The biophysics of marine larval dispersal.*

Pearce, F., 1999. 'Weather warning'. New Sc. 164 (2007): 36–9.

Peck, S.B., 1996. 'Origin and development of an insect fauna on a remote archipelago: The Galálagos Islands, Ecuador.' In Keast, A. and S. Miller, *The origin and evolution of Pacific Island biotas, New Guinea to Eastern Polynesia: patterns and processes.*

Peires, J.B., 1982. *The house of Phalo.*

Perry, T.M., 1963. *Australia's first frontier; the spread of settlement in New South Wales, 1788–1829.*

Peterson, N. and J. Long, 1986. *Australian territorial organisation: a band perspective.*

Pfeffer, P. and J.O. Caldecott, 1986. 'The bearded pig (*Sus barbatus*).' In Phillipps, H., 1990. 'The Stilt, the shrimp, and the scientist'. Aust. Nat. Hist. 23 (4): 322–9. 'East Kalimantan and Sarawak.' Journal of the Malaysian Branch of the Royal Asiatic Society 59: 81–100.

Philander, S.G., 1998. *Is the temperature rising?*

Piddocke, S., 1976. 'The potlatch system.' In A.P. Vayda (ed), *Environment and cultural behaviour: ecological studies in cultural anthropology.*

Pigafetta, A., 1969. *Magellan's voyage: a narrative account of the first circumnavigation.* Ed. by R.A. Skelton.

Popper, W., 1951. *The Cairo Nilometer.*

Potter, L., 1993. 'Banjarese in and beyond Hulu Sengai, South Kalimantan.' In Lindblad, J.T. (ed), *New challenges in the modern economic history of Indonesia.*

Potter, L., 1997. 'A forest product out of control: Gutta percha in Indonesia and the wider Malay world, 1845–1915.' In P. Boomgaard, F. Colombijn and D. Henley (eds), *Paper Landscapes: explorations in the environmental history of Indonesia.*

Potter, L., 1997. 'Where there's smoke there's fire.' Search 28 (10): 307–11.

Pounds, J.A., M.P. Fogden and J.H. Campbell, 1999. 'Biological response to climate change on a tropical mountain.' Nature 398: 611–15.

Pounds, J.A. and M.L. Crump, 1994. 'Amphibian declines and climate disturbance: the case of the golden toad and the harlequin frog.' Cons. Biol 8 (1): 72–85.

Preble, Rev. T.M., 1847. *The Voice of God; or an account of the unparalleled fires, hurricanes, floods and earthquakes, commencing with 1845.*

Predavec, M. and C.R. Dickman, 1994. 'Population dynamics and habitat use of the long-haired rat in southwestern Queensland.' Wildl. Res. 21: 1–10.

Prescott, W.H., 1847. *History of the conquest of Peru.*

Presidency of Madras, 1877. *History of the Madras Famine.*

Priddle, J. et al, 1988. 'Large-scale fluctuations in distribution and abundance of krill – a discussion of possible causes'. In Sahrhage, D. (ed), *Antarctic Ocean and resources variability.*

Queller, D.E., 1997. *The fourth crusade: the conquest of Constantinople, 1201–1204.*

Quinn, D.B., 1974. *England and the discovery of America 1541–1620.*

Quinn, D.B., 1984. *The lost colonists: their fortune and probable fate.*

Quinn, D.B., 1985. *Set fair Roanoke, voyages and colonies, 1584–1606.*

Quinn, W.H. and V.T. Neal, 1987. 'El Niño occurrences over the past four and a half centuries'. J. Geophys. Res. 92: 1449–61.

Quinn, W.H., 1992. 'A study of Southern Oscillation-related climatic for AD 622–1900 incorporating Nile River flood data.' In H. Diaz and V. Markgraf (eds), *El Niño: historical and paleoclimatic aspects of the Southern Oscillation.*

Ransford, O., 1978. *The dark interior.*

Rechten, C., 1985. 'The waved albatross in 1983 El Niño leads to complete breeding failure', In Robinson, G. and E.M. del Pino (eds), *El Niño en las Islas Galapagos: El evento de 1982–1983.*

Redfield, R. and A. Villa Rojas, 1962. *Chan Kom, a Maya village.*

Richardson, J.B. III, 1994. *People of the Andes.*

Richmond, R.H., 1989. 'The effects of the El Niño/Southern Oscillation on the dispersal of corals and other marine organisms.' In Glenn, P. (ed), *Global ecological consequences of the 1982–83 El Niño-Southern Oscillation.*

Riebeeck, J. van, 1952. *Journal of Jan Van Riebeeck.*

Robbins, R.G., 1975. *Famine in Russia 1891–1892.*

Roderick, C., 1991. *Henry Lawson: a Life.*

Rollins, H.B., J.B. Richardson and D.H. Sandweiss, 1986. 'The birth of El Niño: geoarchaeological evidence and implications.' Geoarch. 1 (1): 3–16.

Rolls, E., 1984. *A million wild acres.*

Ropelewski, C.F. and M.S. Halpert, 1987. 'Global and regional scale precipitation patterns associated with the El Niño/Southern Oscillation.' Mon: Weather Rev: 115: 1606–26.

Rougeyron, P., 1846–1849. *Noumea Archives de l'Archevêche.*

Rousey, D.C., 1985. 'Yellow fever and black policemen in Memphis: a post-reconstruction anomaly.' J. Southern Hist. 51 (3): 357–74.

Rowan, M.K. 1967. 'A study of the colies of South Africa.' Ostrich 38: 63–115.

Rowan, R. et al, 1997. 'Landscape ecology of algal symbionts creates variation in episodes of coral bleaching.' Nature 388: 265–9.

Russell, H.C., 1877. *Climate of New South Wales.*

Said, R., 1993. *The river Nile, geology, hydrology and utilization.*

Sakai, S. et al, 1999. 'Plant reproductive phenology over four years including an episode of general flowering in a lowland dipterocarp forest, Sarawak, Malaysia.' Am. J. Bot. 86 (10): 1414–36.

Salomon, F. and G.L. Urioste, 1991. *The Huarochiri Manuscript, a testament of ancient and colonial Andean religion.*

Salstein, D.A. and R.D. Rosen, 1984. 'El Niño and the earth's rotation.' Oceanus 27 (2): 52–57.

Sargent, R.A., 1979. 'The generations of turmoil and stress: a proliferation of states in the northern interlacustrine region, c. 1544–1625.' In Webster, J. B. (ed), *Chronology, migration and drought in Interlacustrine Africa.*

Saunders, M.A., 1998. 'Global warming: the view in 1998.' Report for Benfield Greig Hazard Research Centre.

Schmitt, R.C., 1970. 'Famine mortality in Hawai.' J. Pac. Hist. 5: 109–15.

Schreiber, E.A. and R.W. Schreiber, 1989. 'Insights into seabird ecology from a global "natural experiment".' Nat. Geo. Res. 5(1): 64–81.

Schreiber, E.A., 1994. 'El Niño-Southern Oscillation effects on provisioning and growth in red-tailed tropicbirds.' Colonial Waterbirds 17 (2): 105–19.

Schweithelm, J., 1998. 'The fire this time: an overview of Indonesia's forest fires in 1997–98.' WWF Indonesia discussion paper.

Scott, J.C., 1977. *The moral economy of the peasant; rebellion and subsistence in Southeast Asia.*

Scott, R.F., 1913. *Scott's Last Expedition.*

Scully, W.C., 1898. *Between sun and sand; a tale of an African desert.*

Seavoy, R., 1986. *Famine in peasant societies.*

Sharp, G.D. and D.R. McLain, 1993. 'Fisheries, El Niño-Southern Oscillation and upper-ocean temperature records: an Eastern Pacific example.' Oceanography 6 (1): 13–22.

Shillington, K., 1989. *History of Africa.*

Shimada, I., C.B. Schaff, L. G. Thompson and E. Moseley-Thompson, 1991. 'Cultural impacts of severe droughts in the prehistoric Andes: application of a 1500-year ice core precipitation record.' World Arch. 22 (3): 247–270.

Sinclair, K., 1987. *The Yellow River: a 5000 year journey through China.*

Skinner, J. D., 1993. 'Springbok treks.' Trans. Roy. Soc. S. Afr. 48 (2): 291–305.

Slatin Pasha, R.C., 1990. *Fire and Sword in the Sudan: a personal narrative of fighting and serving the Dervishes, 1879–1895.*

Smith, A., 1776. *The wealth of nations.*

Smith, H.H., 1880. *Brazil, the Amazons and the coast.*

Smith, K. and R. Ward, 1998. *Floods, physical processes and human impacts.*

Smyth, R.B., 1878. *The Aborigines of Victoria: with notes relating to the habits of the natives of other parts of Australia and Tasmania.*

Snow, D.W. and J.B. Nelson, 1984. 'Evolution and adaptations of Galapagos sea-birds.' Biol. J. Linn. Soc. 21: 137–55.

Snow, E., 1962. *The other side of the river: Red China today.*

Solomon, S. and C.R. Stearns, 1999. 'On the role of the weather in the deaths of R.F. Scott and his companions.' Proc. Nat. Acad. Sc. 96 (23) 13012–16.

Soustelle, J., 1964. *The daily life of the Aztecs on the eve of the Spanish conquest.*

Sparrman, A., 1953. *A voyage round the world with Captain James Cook in the HMS Resolution.*

Spear, T.G.P., 1965. *The Oxford history of modern India, 1740–1947.*

Spence, J.D., 1978. *The death of woman Wang; rural life in China in the seventeenth century.*

Sprout, H. and M. Sprout, 1946. *The rise of American naval power, 1776–1918.*

Spruce, R., 1864. *Notes on the valleys of Piura and Chira in Northern Peru.*

Srivastava, H.S., 1968. *The history of Indian famines and development of famine policy, 1858–1918.*

Stahle, D.W. et al, 1998a. 'The lost colony and Jamestown droughts.' Science 280. 564–7.

Stahle, D.W. et al, 1998b. 'Experimental dendroclimatic reconstruction of the Southern Oscillation.' Bull. Am. Met. Soc. 79 (10): 2137–52.

Stein, B., 1998. *A history of India.*

Stevenson, R.L., 1892. *A footnote to history: eight years of trouble in Samoa.*

Stolfi, R.H.S., 1980. 'Chance in history: the Russian winter of 1941–42.' History 65 (214): 214–28.

Strehlow, T.G.H., 1965. 'Culture, social structure and environment in aboriginal central Australia.' In Berndt, R.M and C.H. Berndt (eds), *Aboriginal man in Australia.*

Sturt, C., 1849. *Narrative of an expedition into central Australia during the years 1844, 1845 and 1846.*

Sturt, Capt. C., 1833. *Two expeditions into the interior of Southern Australia during the years 1828, 1829, 1830 and 1831.*

Suttles, W., 1987. *Coast Salish Essays.*

Swan, S.L., 1982. 'Drought and Mexico's struggle for independence.' Environmental Rev. 6 (1): 54–62.

Swan, S.L., 1995. 'La sequía y la disrupción politica.' In Florescano, E. and S. Swan (eds), *Breve historia de la sequía en México.*

Tal, H-T.H., 1983. *Millenarianism and Peasant Politics in Vietnam.*

Tarling, N. (ed), 1992. *Cambridge history of Southeast Asia* Vol. 2.

Tawney, R.H., 1932. *Land and Labour in China.*

Teignmouth, C.J., 1843. *Memoir of the life and correspondence of Lord John Teignmouth.*

Testa, J.W. et. al, 1991. 'Temporal variability in Antarctic marine ecosystems: periodic fluctuations in the phocid seals.' Can. J. Fish. Aquatic Sc. 48 (4): 631–9.

Thayer, V.G. and R.T. Barber, 1984. 'At sea with El Niño.' Nat. Hist. 93: 4–12.

Theal, G.M. (ed), 1902. *Records of southeastern Africa.*

Theal, G.M., 1907. *History of Africa south of the Zambesi* Vol. 2.

Thompson, L.G., E. Moseley-Thompson and B.M. Arnao, 1984. 'El Niño–Southern Oscillation and events recorded in the stratigraphy of the tropical Quelccaya ice cap.' Science 226: 50–53.

Thompson, L.G., E. Moseley-Thompson, W. Dansgaard and P.M. Groots, 1986. 'A 1500 year tropical ice core record of climate: potential relations to man in the Andes.' Science 234: 361–4.

Tiffen, M. and M.R. Mulele, 1994. *The environmental impact of the 1991–2 drought on Zambia.*

Tinker, H.R., 1966. *South Asia: a short history.*

Tinker, H., 1974. *A new system of slavery: the export of Indian labour overseas, 1830–1920.*

Todd, C., 1888. *The Australasian*: 29 Dec: p1456.

Tolcher, H.M., 1986. *Drought or deluge, man in the Cooper's Creek region.*

Totman, C.D., 1993. *Early modern Japan.*

Towle, G.M., 1891. *Magellan.*

Townsend, R.F., 1992. *The Aztecs.*

Trenberth, K.E. and T.J. Hoar, 1996a. 'The 1990–1995 El Niño-Southern Oscillation event: longest on record.' Geophys. Res. Letters 23 (1): 57–60.

Trenberth, K.E. and T.J. Hoar, 1996b. 'El Niño and climate change'. Geophys. Res. Letters 24 (23): 3057–60.

Trenberth, K.E., 1996. 'El Niño-Southern Oscillation.' In Giambelluca, T.W. and A. Henderson-Sellers (eds), *Climate Change: developing southern hemisphere perspectives.*

Trenberth, K.E., G.W. Branstator and P. S. Arkin, 1988. '*Origins of the 1988 North American drought*': Science 242: 1640–5.

Trillmich, F., 1985. 'Drastic effects of El Niño on Galapagos pinnipeds.' Oecologia (Berlin) 67:19–22

Trillmich, F., 1987. 'Seals under the sun.' Nat. Hist. 10/87.

Trillmich, F., 1991. 'El Niño in the Galapagos Islands: A natural experiment.' In Mooney, H.A. et al (eds), *Ecosystem experiments*.

Trillmich, F., 1993. 'Influence of rare ecological events on pinniped social structure and population dynamics.' Symp. Zool. Soc. Lond. 66: 95–114.

Tuan, Y-F., 1980. *Landscapes of Fear*.

Valle, C.A. and M.C. Coulter, 1987. 'Present status of the flightless cormorant, Galapagos penguin and greater flamingo populations in the Galapagos Islands, Ecuador, after the 1982–3 El Niño.' Condor 89: 276–81.

Vance, D.J. and D.J. Staples, 1985 'Factors affecting year-to-year variation in the catch of banana prawns *Penaeus merguiensis* in the Gulf of Carpentaria, Australia.' J. Cons. Int. Expor. Mer. 42: 83–97.

Vancouver, G., 1984. *The voyage of George Vancouver, 1791–1795*.

Vandenbeld, J., 1988. *Nature of Australia*.

Venter, C., 1985. *The Great Trek*.

Vermeij, G.J., 1989. 'An ecological crisis in an evolutionary context: El Niño in the eastern Pacific.' In Glynn, P. (ed), *Global ecological consequences of the 1982–83 El Niño-Southern Oscillation*.

Vidal, J., 1997. 'When the earth caught fire.' *The Guardian*, 8 Nov.

Vogel, C.H., 1989. 'A documentary-derived climatic chronology for southern Africa, 1820–1900.' Clim. Change 14: 291–306.

WWF International, 1997. 'The year the world caught fire.' Int. Discussion Paper.

Waddell, E., 1975. 'How the Enga cope with frost: responses to climatic perturbations in the Central Highlands of New Guinea.' Human Ecology 3 (4): 249–73.

Waddell, E., 1989. 'Observations on the 1972 frosts and subsequent relief programme among the Enga on the Western Highlands.' Mountain Res. and Development 9 (3): 210–223.

Wakeman, F., 1985. *The great enterprise: the Manchu reconstruction of imperial order in seventeenth century China*.

Walford, C., 1879. *The famines of the world: past and present.*

Walker, E.A., 1928. *A history of South Africa.*

Wallace, M.P. and S.A. Temple, 1988. 'Impacts of the 1982–1983 El Niño on population dynamics of Andean condors in Peru.' Biotropica 20 (2): 144–50.

Ward, R., 1980. *The Australian legend.*

Ward, R.G. and M. Brookfield, 1992. 'The dispersal of the coconut: did it float or was it carried to Panama?' J. Biogeog. 19: 467–80.

Watterson, B., 1997. *The Egyptians.*

Webster, J.B., 1980. 'Drought, migration and chronology in the late Malawi littoral.' Trans. J. Hist. 9 (1) 70–90.

Webster, J.B., 1979. 'Noi! Noi! famines as an aid to interlacustrine chronology.' In Webster, J.B. (ed). *Chronology, migration and drought in interlacustrine Africa.*

Webster, J.B., 1983. 'Periodisation in African history, c1050–1840.' J. Gen. Stud. 4 (1): 5–24.

Weiner, J., 1994. *The beak of the finch: evolution in real time.*

Wellington, G.M. and B.C. Victor, 1985. 'El Niño mass coral mortality: a test of resource limitation in a coral reef damselfish population'. Oecologia (Berlin) 68: 15–19.

Wenzel, R.P., 1994. 'A new hantavirus infection in North America.' New England J. Med. 330 (14): 1004–5.

Wheelwright, J., 1994. *Degrees of disaster: Prince William Sound, how nature reels and rebounds.*

Whetton, P. and I. Rutherford, 1994. 'Historical ENSO teleconnections in the eastern hernisphere.' Clim. Change 28: 221–53.

Whetton, P., 1997. 'Floods, droughts and the Southern Oscillation connection.' In Webb, E.K. (ed), *Windows on meteorology: Australian perspective.*

Whetton, P., D. Adamson and M. Williams, 1990. 'Rainfall and river flow variability in Africa, Australia and East Asia linked to El Niño/Southern Oscillation events.' Geol. Soc. Aust. Symp. Proc. 1: 71–82.

White, M.E. and M.W. Downton, 1991. 'The shrimp fishery in the Gulf of Mexico: relation to climatic variability and global atmospheric patterns.' In Glantz, M.H., R.W. Katz and N.

Nicholls (eds), *Teleconnections linking worldwide climate anomalies*.

Wiessner, P., 1980. 'Risk, reciprocity and social influences on !Kung San economics.' In Leacock, E.B. and R.B. Lee, *Politics and history in band societies*.

Wikelski, M. and C. Thom, 2000. 'Marine iguanas shrink to survive El Niño.' Nature 403: 37.

Wikelski, M., 1997. 'Body size and sexual dimorphism in marine iguanas fluctuate as a result of opposing natural and sexual selection: an island comparison.' Evolution 51 (3): 922–36.

Wilkerson, S.J.K., 1994. 'The garden city of El Pital.' Nat. Geo. Res. Expl. 10 (1): 56–71.

Wilkinson, C. et al, 1999. 'Ecological and socioeconomic impacts of 1998 coral mortality in the Indian Ocean: an ENSO impact and a warning of future change?' Ambio 28 (2): 188–94.

Williams, Rev. J., 1837. *A narrative of missionary enterprises in the South Sea islands*.

Williams, S.W., 1899. *The Middle Kingdom*.

Wills, W.J., 1996. *A successful exploration through the interior of Australia, from Melbourne to the Gulf of Carpentaria*.

Wilson, A.N., 1988. *Tolstoy*.

Wilson, D., 1977. *The world encompassed, Drake's great voyages 1577–1580*.

Wilson, D. 1989. *The circumnavigators*.

Wilson, D.J., 1981. 'Of maize and men: a critique of the maritime hypothesis of state origins on the coast of Peru.' Am. Anthrop. 83: 93–120.

Winchester, S., 1997. *The river at the centre of the world: a journey up the Yangtze, and back in Chinese time*.

Wohlt, P., 1989. 'Migration from Yumbisa.' Mountain Res. and Development 9 (3): 210–223.

Woodham-Smith, C., 1980. *The great hunger Ireland 1845–1849*.

Yasuda, M., et al, 1999. 'The mechanism of general flowering in Dipterocarpaceae in the Malay Peninsula.' J. of Trop. Ecol. 15: 437–449.

Zabel, C.J. and S.J. Taggart, 1989. 'Shift in red fox mating system associated with El Niño in the Bering Sea.' Anim. Behav. 38: 830–38.

Zand, K.H., J.A. Videan and I. E. Videan, 1964. *The Eastern Key*.

Zann, R.A. et al, 1995. 'The timing of breeding by zebra finches in relation to rainfall in Central Australia.' Emu 95: 208–22.

Zewde, B., 1976. 'A historical outline of famine in Ethiopia.' In Hussein, A. M. (ed), *Rehab, drought and famine in Ethiopia*.

Ziring, L., 1981. *Iran, Turkey and Afghanistan; a political chronology*.

Index